TRACKING AND PREDICTING THE ATMOSPHERIC DISPERSION OF HAZARDOUS MATERIAL RELEASES

Implications for Homeland Security

Committee on the Atmospheric Dispersion of Hazardous Material Releases
Board on Atmospheric Sciences and Climate
Division on Earth and Life Studies

NATIONAL RESEARCH COUNCIL
OF THE NATIONAL ACADEMIES

THE NATIONAL ACADEMIES PRESS
Washington, D.C.
www.nap.edu

THE NATIONAL ACADEMIES PRESS 500 Fifth Street, NW Washington, DC 20001

NOTICE: The project that is the subject of this report was approved by the Governing Board of the National Research Council, whose members are drawn from the councils of the National Academy of Sciences, the National Academy of Engineering, and the Institute of Medicine. The members of the committee responsible for the report were chosen for their special competences and with regard for appropriate balance.

Support for this project was provided by the National Science Foundation and the National Aeronautics and Space Administration under Grant No. ATM-0135923, and the National Oceanic and Atmospheric Administration under Contract No. 50-DGNA-A-90024. Any opinions, findings, and conclusions or recommendations expressed in this publication are those of the author(s) and do not necessarily reflect the views of the sponsors or their subagencies. Additional funding was provided by the National Research Council.

International Standard Book Number 0-309-08926-3 (Book)
International Standard Book Number 0-309-50935-1 (PDF)

Library of Congress Control Number 2003107398

Additional copies of this report are available from the National Academies Press, 500 Fifth Street, N.W., Lockbox 285, Washington, DC 20055; 800-624-6242 or 202-334-3313 (in the Washington metropolitan area); Internet, http://www.nap.edu.

Cover: (Clockwise from top) The first image is a 3-D representation of tracer plume concentrations from a simulation of a tracer release during URBAN 2000 in Salt Lake City in October 2000. The release location was approximately at the short, black line segment located in the bright red area of the plume. Many complex phenomena related to the interaction of the prevailing southeast wind, the buildings, and the tracer are shown. In particular, there is significant movement of tracer material to the south and to the west in comparison to what would be predicted by a model without explicit representation of buildings. This simulation was done with the HIGRAD CFD model at Los Alamos National Laboratory. The second image is a representation of tracer concentration contours from point measurements made during URBAN 2000 in Salt Lake City. Concentration contour analysis was done by Jerry Allwine at Pacific Northwest National Laboratory. The third image is a simulation of a material release in Portland using the UDM model. The release point is at the red/orange star and the wind is due west. This illustrates the potentially dramatic effect of buildings and street canyons on the behavior of a plume in an urban area. This simulation was done by Los Alamos National Laboratory, and use of the UDM model is courtesy of the Defence Science and Technology Laboratory of the UK Ministry of Defence. The fourth image is a 2-D representation of the same tracer plume concentration described for Image 1.

Copyright 2003 by the National Academy of Sciences. All rights reserved.

Printed in the United States of America

THE NATIONAL ACADEMIES
Advisers to the Nation on Science, Engineering, and Medicine

The **National Academy of Sciences** is a private, nonprofit, self-perpetuating society of distinguished scholars engaged in scientific and engineering research, dedicated to the furtherance of science and technology and to their use for the general welfare. Upon the authority of the charter granted to it by the Congress in 1863, the Academy has a mandate that requires it to advise the federal government on scientific and technical matters. Dr. Bruce M. Alberts is president of the National Academy of Sciences.

The **National Academy of Engineering** was established in 1964, under the charter of the National Academy of Sciences, as a parallel organization of outstanding engineers. It is autonomous in its administration and in the selection of its members, sharing with the National Academy of Sciences the responsibility for advising the federal government. The National Academy of Engineering also sponsors engineering programs aimed at meeting national needs, encourages education and research, and recognizes the superior achievements of engineers. Dr. Wm. A. Wulf is president of the National Academy of Engineering.

The **Institute of Medicine** was established in 1970 by the National Academy of Sciences to secure the services of eminent members of appropriate professions in the examination of policy matters pertaining to the health of the public. The Institute acts under the responsibility given to the National Academy of Sciences by its congressional charter to be an adviser to the federal government and, upon its own initiative, to identify issues of medical care, research, and education. Dr. Harvey V. Fineberg is president of the Institute of Medicine.

The **National Research Council** was organized by the National Academy of Sciences in 1916 to associate the broad community of science and technology with the Academy's purposes of furthering knowledge and advising the federal government. Functioning in accordance with general policies determined by the Academy, the Council has become the principal operating agency of both the National Academy of Sciences and the National Academy of Engineering in providing services to the government, the public, and the scientific and engineering communities. The Council is administered jointly by both Academies and the Institute of Medicine. Dr. Bruce M. Alberts and Dr. Wm. A. Wulf are chair and vice chair, respectively, of the National Research Council.

www.national-academies.org

COMMITTEE ON THE ATMOSPHERIC DISPERSION OF HAZARDOUS MATERIAL RELEASES

ROBERT J. SERAFIN (*chair*), National Center for Atmospheric Research, Boulder, Colorado
ERIC J. BARRON, Pennsylvania State University, University Park
HOWARD B. BLUESTEIN, University of Oklahoma, Norman
STEVEN F. CLIFFORD, University of Colorado, CIRES, Boulder
LEWIS M. DUNCAN, Dartmouth College, Hanover, New Hampshire
MARGARET A. LEMONE, National Center for Atmospheric Research, Boulder, Colorado
DAVID E. NEFF, Colorado State University, Fort Collins
WILLIAM E. ODOM, Hudson Institute, Washington, D.C.
GENE J. PFEFFER, Ridgefield Consulting, Colorado Springs, Colorado
KARL K. TUREKIAN, Yale University, New Haven, Connecticut
THOMAS J. WARNER, University of Colorado, Boulder
JOHN C. WYNGAARD, Pennsylvania State University, University Park

NRC Staff

LAURIE GELLER, Study Director
VAUGHAN C. TUREKIAN, Program Officer (through August 2002)
DIANE GUSTAFSON, Administrative Associate
JULIE DEMUTH, Research Associate

BOARD ON ATMOSPHERIC SCIENCES AND CLIMATE

ERIC J. BARRON (*chair*), Pennsylvania State University, University Park
RAYMOND J. BAN, The Weather Channel, Inc., Atlanta, Georgia
ROBERT C. BEARDSLEY, Woods Hole Oceanographic Institution, Massachusetts
ROSINA M. BIERBAUM, University of Michigan, Ann Arbor
HOWARD B. BLUESTEIN, University of Oklahoma, Norman
RAFAEL L. BRAS, Massachusetts Institute of Technology, Cambridge
STEVEN F. CLIFFORD, University of Colorado, CIRES, Boulder
CASSANDRA G. FESEN, Dartmouth College, Hanover, New Hampshire
GEORGE L. FREDERICK, Vaisala Inc., Boulder, Colorado
JUDITH L. LEAN, Naval Research Laboratory, Washington, D.C.
MARGARET A. LEMONE, National Center for Atmospheric Research, Boulder, Colorado
MARIO J. MOLINA, Massachusetts Institute of Technology, Cambridge
MICHAEL J. PRATHER, University of California, Irvine
WILLIAM J. RANDEL, National Center for Atmospheric Research, Boulder, Colorado
RICHARD D. ROSEN, Atmospheric & Environmental Research, Inc., Lexington, Massachusetts
THOMAS F. TASCIONE, Sterling Software, Inc., Bellevue, Nebraska
JOHN C. WYNGAARD, Pennsylvania State University, University Park

Ex Officio Members

EUGENE M. RASMUSSON, University of Maryland, College Park
ERIC F. WOOD, Princeton University, New Jersey

NRC Staff

CHRIS ELFRING, Director
ELBERT W. (JOE) FRIDAY, JR., Senior Scholar
LAURIE S. GELLER, Senior Program Officer
AMANDA STAUDT, Program Officer
JULIE DEMUTH, Research Associate
ELIZABETH A. GALINIS, Project Assistant
ROB GREENWAY, Project Assistant
DIANE L. GUSTAFSON, Administrative Associate
ROBIN A. MORRIS, Financial Officer

Preface

In the wake of the terrorist attacks of September 11, 2001, the National Academies launched a major new initiative to provide guidance to the federal government on scientific and technical matters related to counterterrorism and homeland security.[1] All of the boards within the National Academies were asked to consider how their particular research communities could contribute to this effort. The Board on Atmospheric Sciences and Climate (BASC) discussed this matter at its autumn 2001 meeting and proposed the idea for the workshop that is described in this report.

There is growing concern that future terrorist activities may involve the release of chemical or biological weapons or the detonation of "dirty bombs" that release radioactive material. Atmospheric observations and models can be used to track a hazardous release and to forecast how a plume of hazardous material may spread. Emergency responders can use this information to identify affected locations and make life-saving decisions about evacuating or sheltering endangered populations. The BASC members agreed that there was a great need to critically examine the observational and modeling tools used for tracking the atmospheric dispersion of chemical, biological, or nuclear (C/B/N) agents and to assess the value of dispersion forecasts for providing useful information to emergency responders and the general public.

To address these issues, a steering committee was convened that included several members of the BASC and a number of additional people chosen to augment the group's expertise. The steering committee held an initial planning meeting on May 8-9, 2002, in Washington, D.C., and the workshop was held on July 22-24, 2002, in Woods Hole, Massachusetts. The charge to the committee was to organize a workshop that addressed the following tasks:

---

[1] A centerpiece of this effort is the report *Making the Nation Safer: The Role of Science and Technology in Countering Terrorism* (NRC, 2002c). That report was produced by a parent committee (chaired by Lewis Branscomb and Richard Klausner) that synthesized the analysis of eight subpanels.

- Review the current suite of atmospheric models that are used in characterizing atmospheric dispersion and examine how these models are applied operationally for emergency response efforts.
- Identify deficiencies in the models that limit their effectiveness and breadth of application; assess the research and development needed to enhance the effectiveness and operational use of these models in emergency situations.
- Determine the observational data needed to initialize, test, and use these models effectively, and identify ways that other environmental measurements can complement these models to provide additional and more accurate information.

This activity focused on tracking terrorist releases of C/B/N agents (primarily focusing on local- or regional-scale dispersion), but it should be noted that many of the issues raised in this context are applicable to tracking other hazardous materials dispersed through the atmosphere, such as air pollution, smoke from forest fires, and industrial chemical spills. Note also that the workshop participants did recognize C/B/N sensors as critically important components of a dispersion tracking and forecasting system. An examination of sensor technologies can be found in another recent National Research Council report (NRC, 2002a).

There are dozens of dispersion models in use as operational or research tools. We did not attempt to carry out a comprehensive model analysis or intercomparison, but instead, we examined a small subset of modeling systems (primarily those used by national agencies) that represent a range of capabilities and applications. The models chosen for discussion here do not represent the committee's judgment about the "best" systems. Recently, the Office of the Federal Coordinator for Meteorology and the Department of Defense each carried out a comprehensive survey of available dispersion models and their capabilities (although both assessments were largely qualitative in nature). This National Academies' activity is aimed at complementing the governmental activities by providing an independent forum for assessing our nation's current capabilities and needs. This activity was initiated internally and supported through National Academies' endowment and BASC core funds (received from the National Science Foundation, National Oceanic and Atmospheric Administration, and National Aeronautics and Space Administration).

This report provides a summary of the discussions that took place at the workshop. It is organized along the same lines as the workshop itself, divided into three main topics: (1) information requirements of the emergency response community, (2) observational capabilities and needs, and (3) modeling capabilities and needs. The workshop focused primarily on informal discussion among the participants, but it also included a few presentations to provide background information and context for the participants. The appendixes of this report include a summary of several of these presentations.

Robert J. Serafin, *Chair*

Acknowledgments

This report has been reviewed in draft form by individuals chosen for their diverse perspectives and technical expertise, in accordance with procedures approved by the National Research Council's Report Review Committee. The purpose of this independent review is to provide candid and critical comments that will assist the institution in making its published report as sound as possible and to ensure that the report meets institutional standards for objectivity, evidence, and responsiveness to the study charge. The review comments and draft manuscript remain confidential to protect the integrity of the deliberative process. We wish to thank the following individuals for their review of this report:

    Donald E. Aylor, Connecticut Agricultural Experimentation Center
    David P. Bacon, Science Applications International Corporation
    Donald L. Ermak, Lawrence Livermore National Laboratory
    Jack Fellows, University Corporation for Atmospheric Research
    George L. Frederick, Vaisala Inc.
    Robert G. Hendrickson, Oregon Health and Science University
    Andrew Majda, New York University
    Clifford F. Mass, University of Washington
    Stephen J. McGrail, Massachusetts Emergency Management Agency
    Frances Edwards-Winslow, San Jose Office of Emergency Services

Although the reviewers listed above have provided constructive comments and suggestions, they were not asked to endorse the report's conclusions or recommendations, nor did they see the final draft of the report before its release. The review of this report was overseen by John F. Ahearne, Sigma Xi, The Scientific Research Society, and Charles E. Kolb, Aerodyne Research, Inc. Appointed by the National Research Council, they were responsible for making certain that an independent examination of this report was carried out in accordance with institutional procedures and that all review comments were carefully considered. Responsibility for the final content of this report rests entirely with the authoring committee and the institution.

Contents

EXECUTIVE SUMMARY		1
1	**INTRODUCTION**	8
2	**USER NEEDS**	11
	Preparedness, 11	
	Response, 12	
	Recovery and Analysis, 17	
	Key Findings and Recommendations, 18	
3	**OBSERVATIONAL CAPABILITIES AND NEEDS**	19
	Plume Identification, 20	
	Wind—Local Flows, 20	
	Depth and Intensity of Turbulent Layers, 26	
	Deposition and Degradation, 27	
	Key Findings and Recommendations, 29	
4	**DISPERSION MODELING: APPLICATION TO C/B/N RELEASES**	33
	Categories of Dispersion Models, 34	
	Interpreting and Evaluating Dispersion Model Outputs, 35	
	Overview of C/B/N Dispersion Modeling Systems, 40	
	Review of Selected C/B/N Dispersion Modeling Systems, 44	
	Discussion of C/B/N Modeling Systems, 48	
	Key Findings and Recommendations, 51	
REFERENCES		55
ACRONYMS AND ABBREVIATIONS		57
COMMITTEE BIOGRAPHIES		60

APPENDIXES 63

A Workshop Agenda and Participant List, 65
B Overview of Atmospheric Transport and Dispersion Modeling, 69
C Meteorological Observing Systems for Tracking and Modeling C/B/N Plumes, 72
D Scientific and Technical Information Needs of Emergency First Responders, 78
E Ensemble Simulations with Coupled Atmospheric Dynamic and Dispersion Models: Illustrating Uncertainties in Dosage Simulations, 80
F Modeling Studies of the Dispersion of Smoke Plumes from the World Trade Center Fires, 85
G Use of Atmospheric Models in Response to the Chernobyl Disaster, 87
H Preparatory Exercises at the Salt Lake City Olympics, 89
I URBAN 2000 Overview, 91

Executive Summary

The National Academies workshop "Tracking and Predicting the Dispersion of Hazardous Agents" brought together atmospheric scientists from academia, government laboratories, and the private sector; emergency management officials and first responders; and experts in national security, risk communication, and other relevant fields. Workshop participants examined how meteorological observations and dispersion models can be used by emergency managers in the context of an atmospheric release of hazardous chemical, biological, or nuclear (C/B/N) agents. It was found that atmospheric observational and modeling tools can contribute substantively to preparation and planning for possible future events, to emergency response in the minutes to hours after an event occurs, and to the post-event recovery and analysis. Existing capabilities generally are useful, but emergency responders have a number of observational and modeling needs that are not well satisfied by existing services. Although it may never be possible to provide a "perfect" atmospheric dispersion prediction for any individual hazardous release, the committee believes that with more effective application of available tools and development of new technologies and capabilities, the atmospheric science community could play a larger role in addressing this critical national security concern.

The organizing committee extracted a number of important lessons from the workshop discussions and, in its subsequent deliberations, identified the following as key findings and recommendations.

MEETING THE NEEDS OF EMERGENCY RESPONDERS

Atmospheric observations and dispersion models must interface seamlessly with the needs of emergency responders. Emergency response managers would benefit from training that conveys the strengths and weaknesses of existing observational and dispersion modeling tools and the situations under which various types of tools perform best. Conversely, dispersion modelers and meteorologists would benefit from learning how nowcasts and forecasts are used in emergency response situations. **"Tabletop" (i.e., roundtable discussion and planning) event simulation exercises should be convened** regularly to bring together emergency response teams and members of the atmospheric modeling and observational communities to help establish and exercise a common set of data interface and decision support protocols.

Emergency responders face a confusing array of seemingly competitive atmospheric transport model systems supported by various agencies, and in many cases, they do not have a clear understanding of where to turn for immediate assistance. **A single federal point of contact should be established (such as a 1-800 phone number) that could be used to connect emergency responders across the country to appropriate dispersion modeling centers for immediate assistance.**

Emergency managers need a realistic understanding of the bounds on the uncertainties of dispersion model predictions. Dispersion model predictions of the concentrations for a given release need to be accompanied by a prediction of the event-to-event variability in that situation. **Dispersion modelers should use ensemble modeling or other approaches that quantify not only the average downwind concentration distribution in a given situation (which is interpretable as the most likely outcome) but also the event-to-event variability to be expected. The specific formats of the information presented should be developed in close collaboration with users of this information.**

ENHANCING OBSERVATIONAL RESOURCES

The most basic observations required for tracking and predicting the dispersion of a hazardous agent include identification of the plume, characterization of low-level winds (to follow the plume trajectory), characterization of the depth and intensity of the turbulent layers through which the plume moves (to estimate plume spread), and identification of areas of potential agent degradation and dry or wet deposition.

The current array of surface observational systems needs to be better used and enhanced. Many surface stations are poorly exposed and have limited instrument quality control, and instrument locations are not necessarily optimal for model initialization or identification of local flows. Furthermore, it often is difficult to obtain the data from multiple observational arrays, especially in real time. **A comprehensive survey of the capabilities and limitations of existing observational networks should be conducted, followed by action to improve these networks and access to them, especially around more vulnerable areas.**

Doppler radar systems can be useful for estimating boundary layer winds, monitoring precipitation, and possibly tracking some C/B/N plumes. NRC (2002b) recommended evaluating the potential for supplementing the current Doppler radar network with subnetworks of short-range, short-wavelength radars. This would enable better estimates and coverage of low-level winds, increase the likelihood of detecting C/B/N plumes, and improve precipitation (and hence wet deposition) estimates. **The committee supports this recommendation and further recommends that the design and data collection strategy of this radar network be optimized to include providing information for supporting response to a C/B/N release.**

Radar wind profilers and radio acoustic sounding system profilers, which measure variations of the horizontal wind and temperature, respectively, with height and enable identification of turbulent layers, provide important information for response to C/B/N attacks and are relatively inexpensive and easy to maintain. **Wind and temperature profilers should become an integral part of regional and local fixed-observational networks.**

Mobile observational platforms can provide valuable information and fulfill multiple needs in the first minutes to hours after a hazardous release. Unmanned aerial vehicles (UAVs) can be used to measure wind and temperature profiles and to characterize turbulence where other platforms cannot easily reach. Mobile lidars and radars can, in some contexts, be used for plume tracking and wind field characterization. However, civilian instruments currently are available only for research use. **There should be continued development of portable scanning lidars and radars on airborne and surface-mobile platforms for research, and plans should be developed to make such instruments rapidly available for effective, timely use in vulnerable areas.**

Local topography and the built environment lead to local wind patterns that can carry contaminants in unexpected directions. Observational networks must represent these local flows as faithfully as possible. Improvements in these networks can be achieved through routine data monitoring and comparison of observed flows with local- to regional-scale model simulations and through numerical modeling, including observing system simulation experiments. Studies should be performed over a range of weather situations and for both daytime and nighttime conditions. Such exercises will educate meteorologists about local flows and model capabilities; the resulting knowledge of what to believe when observational data and models convey different messages is vital in response to an emergency situation. **Efforts should be made to systematically characterize local-scale windflow patterns (over the full diurnal cycle) in areas deemed to be potential terrorist targets with the goals of optimizing fixed observations and educating those involved in developing dispersion forecasts about local flows and model strengths and weaknesses.**

Focused field exercises are needed to understand the behavior of modeled transport and dispersion in different weather regimes and C/B/N release scenarios, particularly for nocturnal conditions. It is not practical to verify dispersion and transport models for every area with comprehensive field programs, but for an appropriate range of meteorological conditions, physical modeling in a wind tunnel could assist in dispersion model evaluation and threat assessment. In addition, field programs conducted for other purposes, such as improvement of weather forecasting or understanding boundary layer turbulence, also can be useful. **There should be continued field programs focused on C/B/N release issues, and datasets from field programs with a C/B/N or related focus should be made available for testing and development of dispersion and mesoscale transport models.**

Some of the actions recommended above (i.e., enhancing fixed observing arrays, optimizing placement of surface stations and wind profilers, developing and deploying portable scanning lidars, UAVs, and radars) will be costly. **There should be priori-**

tization of such actions based on identifying areas with the greatest need (e.g., highest population concentration, most complex flow, greatest likelihood for a terrorist attack, most vulnerable facilities). Every effort should be made to utilize such instrumentation for other (hazardous and non-hazardous) applications (e.g., to enhance air pollution monitoring, optimize agricultural practices, aid in severe-storm forecasting and highway network safety), thus sharing the costs and ensuring that the array will be continuously used, maintained, evaluated, and quality controlled.

STRENGTHENING MODELING CAPABILITIES AND APPLICABILITY TO EMERGENCY RESPONSE

For purposes of threat assessment, preparation, and training, existing dispersion models meet some needs of the emergency response community. In the case of actual emergencies, the needs of emergency management may not be well satisfied by existing models. In particular, single-event uncertainties in atmospheric dispersion models are not well bounded, and current models are not well designed for complex natural topographies or built urban environments.

Most available atmospheric dispersion models predict only the ensemble-average concentration (that is, the average over a large number of realizations of a given dispersion situation). New approaches are needed for modeling a single hazardous release.

Dispersion models used for emergency planning and response should provide confidence estimates that prescribed concentrations will not be exceeded outside of predicted hazard zones. This requires that models provide some measure of the possible variability in a given situation.

Different dispersion modeling methodologies are required in the preparedness, response, and recovery stages of C/B/N events. For the preparedness stage, an accurate model capable of providing confidence-level estimates is desired, but model execution time is not important. For the response stage, accuracy can be compromised to obtain timely predictions, but the dispersion model must still provide confidence-level estimates. For the recovery stage, model execution time is not important, but accurate model reconstruction of the plume concentration distribution over time is desired. In order to use a dispersion model's predictions effectively during the early response phase, the wind field and other conditions at the site of the release must be available in near real time and a short model execution time is essential. The most appropriate dispersion model for any given scenario may depend on the quantity, toxicity, and persistence of the hazardous agent; thus, it is critical that source identification be as rapid as possible.

The committee's review of selected existing dispersion modeling systems determined that no one system had all the features that the committee deemed critical: confidence estimates for the predicted dosages, accommodation of urban and complex topography, short execution time urban models for the response phase, and accurate though slower models for the preparedness and recovery phases. Better integration between existing and future modeling systems could supply all of these critical features.

The "unpairing" of concentration predictions and observations in time and space (commonly done with continuous sources in air quality applications) is inappropriate when evaluating dispersion model performance in episodic releases. Evaluation techniques based on more advanced probabilistic methods need to be developed. Toward that end, existing dispersion models should identify the type of averaging (ensemble, time and space) inherent in their modeling methodology, both in the wind field formulation and in the treatment of dispersion. The reliability of existing and future dispersion modeling systems should be evaluated against field and laboratory measurements for potential C/B/N event scenarios. If predicted confidence limits are found to be unacceptable, then empirical corrections should be applied to model outputs so as not to place emergency personnel in harm's way. Increasing the density of the wind measurements in a plume's domain will potentially reduce uncertainty, thus reducing the predicted extent of the hazard without compromising confidence.

Meteorological observations are a critical element of dispersion modeling. Observational technologies have been evolving rapidly in recent decades, and the committee identified many existing measurement technologies that have not been fully exploited through data assimilation. Model operators and developers would benefit from broader interaction with the meteorological community to take advantage of leading-edge research in data assimilation, quantitative precipitation forecasting, short-range numerical weather prediction, and high-resolution forecasting initialized with radar data. Likewise, observational research programs studying issues such as weather prediction, properties of boundary layer turbulence, and air pollution transport should be viewed as targets of opportunity for testing and evaluating dispersion models.

Priorities for improving modeling capabilities include the following:

- **New dispersion modeling constructs need to be further explored and possibly adapted for operational use in urban settings. This includes advanced, short execution time models, slower but more accurate computational fluid dynamics and large-eddy simulation models, and models with adaptive grids.**
- **Techniques must be developed for constructing ensembles of model solutions on the urban scale so that probabilistic rather than deterministic information can be provided to emergency managers. It will be necessary to quantify the level of confidence as a function of the number of ensemble members, which in turn, will have implications for the computational power required.**
- **It is necessary to learn how to more effectively assimilate into models an appropriate range of meteorological data (e.g., wind, temperature, and moisture data) from observing systems as well as real-time data from C/B/N sensors, especially as the quality and availability of these data increase. It also is important to effectively couple dispersion models with appropriate source characterization models.**
- **Urban field programs and wind-tunnel urban simulations should be conducted to allow for the testing, evaluation, and development of existing and new modeling systems (both meteorological and dispersion models). Developing an appropriate experimental design for such studies is a critical task that itself will require careful evaluation.**

- The bulk effects of urban surfaces on the surface energy, moisture, and momentum are not well accounted for in most meteorological models. Existing development work in this area should be enhanced and the improved modeling techniques adopted more widely.
- Urban building and topography three-dimensional databases need to be developed and maintained for use in numerical and wind-tunnel dispersion simulations.

MANAGEMENT AND COORDINATION NEEDS

There is a wide array of federal agencies that operate dispersion modeling systems, including the Department of Commerce–National Oceanic and Atmospheric Administration, Department of Defense, Department of Energy, Environmental Protection Agency, Federal Emergency Management Agency, and Nuclear Regulatory Commission, along with numerous academic and private sector research groups that contribute to these federal efforts. In addition, it must be recognized that the new Department of Homeland Security, established in January 2003, may eventually augment or subsume some of the activities and responsibilities currently residing in these other federal agencies. At the present time, however, it is not known to the committee what specific organizational plans are being considered.

Given the ambiguity of this situation and the limited time and resources available to examine these management-related issues, the committee felt that it was not appropriate to make specific suggestions about agency leadership responsibilities for the various activities recommended in this report. The committee emphasizes, however, that a carefully crafted management strategy, with clear lines of responsibility and authority, is essential for ensuring further progress in the development and ongoing operation of dispersion modeling systems. There is a clear need for more central coordination among the various federal agencies currently involved and among the relevant players at the local, regional, and national levels.

Each of the agencies mentioned above has developed its own "customer base" and areas of strength and specialization; thus, it seems likely that some form of distributed responsibility will continue to be the most effective organizational strategy. However, a strong center of coordination is needed to ensure that the necessary research and development work is carried out and that emergency responders have unambiguous guidance as to where to turn for help.

A nationally coordinated effort should be established to foster support and systematic evaluation of existing models and research and development of new modeling approaches, undertaken in collaboration with the broader meteorological community. The Office of the Federal Coordinator for Meteorology, which recently organized a review of U.S. dispersion modeling capabilities, could provide valuable input as to which agency(ies) is best suited to oversee this coordinated effort.

In at least one large urban area, a fully operational dispersion tracking and forecasting system should be established—that is, a comprehensive system for

collecting relevant meteorological and C/B/N sensor data, assimilating this information into a dispersion model, and maintaining the expertise and organizational capacity to provide immediate model forecasts on a full-time basis. If possible, a few such systems should be established and evaluated for different types of urban areas (e.g., coastal versus continental cities, low-latitude versus high-latitude cities). Such systems can be used as test beds for gaining understanding of model capabilities and limitations, and their use should not be limited to emergency situations. These observational and modeling tools could have multiple applications, which would help justify costs and ensure that the systems are frequently used, maintained, evaluated, and quality controlled.

There is a wealth of knowledge about meteorological and dispersion models residing in universities, National Weather Service Weather Forecast Offices, and private sector facilities throughout the nation. These sources of expertise, together with the existing programs in several national laboratories and military facilities, should be integral components of the coordinated national effort recommended above, to assist with developing local and regional models that are optimized for the topography and seasonal weather patterns in vulnerable areas. At the most basic level, this integration can be implemented via collaborative research and development efforts.

1

Introduction

Today, in our post-Cold War world, terrorism represents the single most prominent threat to global security, economy, and social order. As expanding societal globalization has brought different cultures and ideas into closer contact and occasional conflict, technological globalization also has increased societal access to weapons of mass destruction. Terrorism, including the threatened use of weapons of mass destruction against innocent civilian populations, has been adopted as a key instrument of asymmetric conflict between groups or nations of disparate political and military standing in the world. Given the complex and often ambiguous purposes motivating these acts, the anticipation, preparedness, prevention, and response to terrorism is exceedingly difficult, requiring a dedicated refocusing of our national security efforts. Terrorism today spans the use of traditional methods of warfare such as conventional explosives to the emerging possibility of the use of weapons of mass destruction, including chemical, biological, and nuclear (C/B/N) agents. Particularly for these weapons of mass destruction, anticipation and assessment of the dispersal of harmful agents is a critical element of our counterterrorism preparedness and response.

Airborne releases of hazardous agents have been a principal concern of communities and emergency managers (Appendix D). Communities have prepared themselves to deal with accidental releases from industrial sites, energy facilities, and vehicles transporting hazardous materials. The military has been concerned with chemical and biological warfare as well as the potential for tactical nuclear weapons in the battlefield.

Dispersion models are important tools for dealing with all of these issues, and observations ranging from direct visual sightings to sophisticated sensor measurements, provide essential input for these modeling systems.[1] The workshop that is the focus of this report examined the application of such observations and dispersion models in a wide

---

[1] The requirements of a truly comprehensive operational system go well beyond the technical modeling and observational tools discussed here. For instance, such a system requires appropriate expertise and capabilities for effective communications, data acquisition, and "user-friendly" product display. Although these factors are all vitally important, they are beyond the scope of this study.

variety of contexts, such as prediction of the global transport of radioactive material from the Chernobyl accident (Appendix G); dispersion of a possible hazardous agent release (Appendix E); preparatory planning for the 2002 Winter Olympics in Salt Lake City (Appendix H); analysis of the dispersion of smoke plumes from the World Trade Center disaster (Appendix F); and air quality research and prediction.

The basic components of a dispersion modeling system are illustrated in Figure 1.1. As is evident from the figure, a comprehensive model takes into account the nature of the material released, local topography, and meteorological and atmospheric data; and from this information is derived some form of risk parameters. Current available modeling systems range from the relatively simple to the highly complex (for example, a sophisticated model may contain its own meteorological prediction component and interactive atmospheric chemistry).

In order to determine how dispersion models can be applied most effectively, it is important to identify the end users and assess their needs. In doing so, it is useful to classify emergency response activities into the phases of preparedness, response, and recovery and analysis. As discussed in the subsequent chapters of this report, for each of these phases, emergency responders have different information needs and there are different opportunities for utilizing atmospheric observations and models.

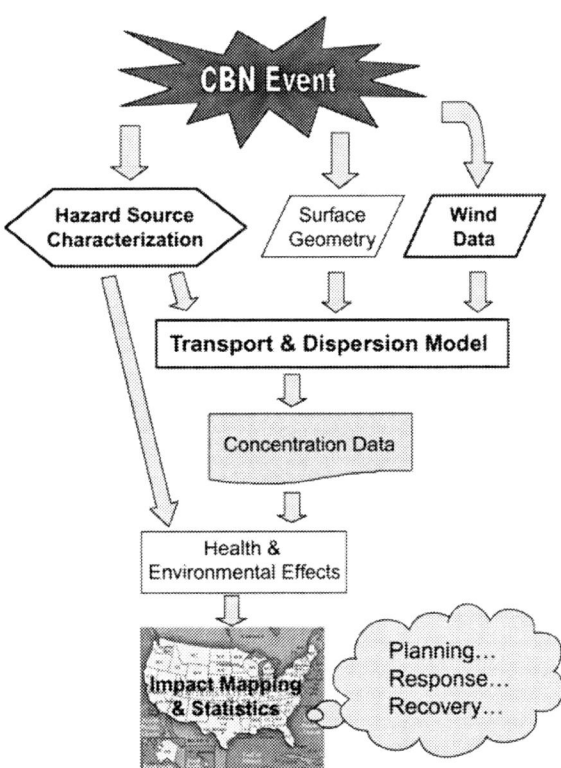

FIGURE 1.1 Chemical, biological, and nuclear event modeling system.

Anticipating and responding to terrorist attacks is extremely challenging because the possible scenarios of timing, location, and method of attack essentially are infinite. In many cases, the exact source location may not be known initially (e.g., it could be an instantaneous release or be distributed temporally and geographically; the source could be ground-based or airborne), and the nature of the substance released may also be unknown initially. Thus, dispersion tracking and forecasting systems must be capable of providing useful information even in the absence of some basic input information.

Much can be done with existing resources to strengthen observational and modeling capabilities for tracking hazardous releases, such as better use of existing local observational networks and better exploitation of existing models. However, additional resources likely will be required by many communities for the development and implementation of improved observing systems and higher-resolution models. Because the terrorist threat probability is small, many communities might find it difficult to justify the investments needed. However, as discussed in this report, robust observing systems and high-resolution atmospheric modeling systems can support many other important functions such as local weather warnings, air quality forecasting, and transportation system management. The combined benefits thus are likely to justify the investments.

These issues are discussed in greater depth in the following chapters. Chapter 2 examines the information needs of emergency responders in the preparedness, response, and recovery and analysis phases of a hazardous release. Chapter 3 examines the role of atmospheric observations in tracking and predicting the dispersion of hazardous agents, including an assessment of our current observational capabilities and needs for improvement. Chapter 4 contains an overview of the capabilities and limitations of the various types of dispersion models in use today. In each chapter the committee identifies a number of priority findings and recommendations that emerged from discussions among the workshop participants.

2

User Needs

This report assesses the needs for improving our ability to track and predict the atmospheric transport of chemical, biological, or nuclear (C/B/N) atmospheric releases. These needs are defined in terms of the various communities who must respond to such threats and their counterterrorism objectives and decision-support time frames. Different user communities establish and prioritize their needs differently. By identifying end user requirements, the committee has attempted to focus on the practical application and implementation opportunities for atmospheric modeling and observational tools. The broad range of counterterrorism activities is divided into the areas of *preparedness* (which, in turn, includes *intelligence and threat assessment, preparedness planning, prevention and protection*), *response*, and *recovery and analysis*. Each of these stages places a different set of constraints and requirements on observational and modeling needs (Appendix D). Response and recovery needs are further subdivided according to the diversity of responders, their particular responsibilities, and the time scales associated with their various roles.

PREPAREDNESS

Intelligence and threat assessment involves consideration of the capabilities and attack risks linked with any potential terrorist organization, from individuals acting alone to organized groups or even hostile nation-states. Atmospheric transport modeling may contribute in several ways to this expansive effort. The historical precedent of nuclear weapons test monitoring attests to the usefulness of accurate atmospheric modeling studies as a means of retracing the transport of airborne C/B/N agents. Transport modeling may also assist in determining sensor sensitivity and sampling requirements as well as preferential locations for monitoring (either systematically for wide coverage or specifically for suspected terrorist activities). Similarly, such models may be used to assess the risk associated with any number of hypothetical threat scenarios against assumed targets.

Atmospheric transport modeling tools can be used to help determine the time, location, and magnitude of releases after they have occurred. An example of this type of event was the tracking of radioisotopes released from the Chernobyl reactor accident. For these "hindcast" activities, particularly for cases in which extended time is available for after-the-fact analysis, existing large-scale transport models have provided useful support to the intelligence community. However, improved atmospheric dispersion modeling could contribute substantively to the design of enhanced monitoring systems, for instance, to help determine requirements for monitor location and spacing and sensor measurement sensitivities. Assessment of conjectured threats against known potential targets (such as nuclear power plants) also seems to be served satisfactorily by existing atmospheric dispersion models. For these cases, predictions of average atmospheric behavior and likely variations around mean dispersion seem adequate for general threat assessment and training purposes. However, as the need for higher temporal and spatial resolution mapping becomes greater—for example, with regard to threat assessment in urban environments and complex topographies—current transport models are not yet sufficiently useful. Furthermore, given that local-scale transport is affected by dynamic weather conditions, such models require continuous updating with observations or output from meteorological models.

Preparedness planning is a natural extension of counterterrorism threat assessment, and it complements existing emergency planning for accidental atmospheric releases of harmful agents. This particularly is the case for facilities such as petrochemical and nuclear plants that are known to be potential sites of hazardous releases and that may also be terrorist targets. Emergency responders generally have well-established plans and contingency options for reacting quickly to events involving atmospheric releases (whether accidental or purposely induced) from such pre-identified facilities. In many cases, they have trained regularly against such threats. Existing atmospheric transport models appear to be useful for site-specific planning and training needs and likewise for event-specific preparation and planning activities, such as those associated with major entertainment, sports, or other public events (e.g., the Super Bowl, a presidential inauguration).

Protection and prevention generally involve the anticipation and interdiction of suspected terrorist activity by responsible authorities before a terrorist attack occurs and, specifically, before the release of C/B/N agents into the atmosphere. Although successful interdiction implies that an atmospheric release of hazardous material has been avoided, atmospheric transport modeling can and has been used to assist in decision making for the allocation of monitoring resources and deployment of field personnel. For example, during the Salt Lake City Olympics (Appendix H), prevailing weather patterns and predicted atmospheric transport effects were used by protection forces to identify areas of heightened vulnerability or risk and, correspondingly, to help allocate available monitoring resources for maximum coverage and effectiveness. While improved transport modeling would be useful in the case of real emergency events, existing models have proven useful for satisfying these preventive resource-allocation and training needs.

RESPONSE

Once hazardous agents have been released into the atmosphere, a series of emergency response actions will occur, carried out by a variety of specialized emergency response personnel working in concert across several overlapping time scales. Each of these users and time scales places different needs-based requirements on tools for tracking the atmospheric release. For the purpose of providing an assessment framework, actions and user needs are defined on three time scales:

1. Immediate first response (0–2 hours)
2. Early response (generally 2–12 hours)
3. Sustained response support (generally greater than 12 hours)

Response to events also is affected by knowledge of the release source term; for instance, one may have a likely known source agent (such as a nuclear power plant), an unknown source term (such as an undetermined biotoxin release), or a quasi-known source (such as a chemical explosion with visible plume but of uncertain or mixed composition).

"First responder" is a term generally used to describe the fire and rescue, medical services, and law enforcement personnel responding to an emergency over the first several hours (Appendix D). For the purposes of this study, the committee defines first responders as those individuals who are first to report and arrive at the scene of an emergency, often within minutes after the events occur. These individuals frequently will be the ones who report the emergency to local and state emergency response managers, provide an initial assessment of its nature and magnitude, and direct short-term response reaction over the first few tens of minutes of an event. Their highest priority is to protect the public and to care for the injured. Beyond whatever benefits might be derived from preparatory training exercises, there may be little opportunity for atmospheric dispersion modeling to assist in meeting first responders' needs in the immediate aftermath of an actual terrorist attack. In contrast, real-time observations of wind, precipitation, and so forth, may play a major role in immediate decision-making.

Dispersion modelers must understand the role and capabilities of these first responders; they serve as the initial data collection interface on what has happened, often being asked to provide subjective characterization of the release events so as to best determine follow-on emergency responses over the next few hours. Their limited descriptive input may be the only information available for the first quick-look atmospheric model assessment of likely event consequences.

The early emergency response team will move into action upon receiving initial reports of an event (or a series of related events). This response team may be part of larger emergency management teams that are state, county, or municipality based, depending on the event location. Emergency response protocol establishes the official primacy of local authorities in dealing with such emergencies, although state, regional, and federal resources may be actively engaged in providing various degrees of supplemental support. The experience and training of these early emergency response teams is especially crucial during these chaotic first few hours following a release. In larger population centers, a member of the emergency response team likely will have some level of experience and capability in using very simple transport modeling tools (such as CAMEO/ALOHA [Computer-Aided Management of Emergency Operations–Areal Locations of Hazardous Atmospheres], discussed in Chapter 4). In more rural or less prepared locales or for the use of more sophisticated models, emergency response teams may need more complete advisory support from a national or regional atmospheric modeling center.

The initial response plan over the first several hours of an event typically will include execution of a quick-look atmospheric transport model prediction. This model may have very limited access to real-time atmospheric data and information about the hazard source (in terms of injection dynamics, aerosol size and composition, and potential lethal dosage[1]). Over the next few hours, as additional real-time data and source term information become available, modeling predictions will become increasingly accurate and specific.

It is essential that the atmospheric modeling results support the decision-making needs of this early responder community (Appendix D). Within the first few tens of minutes to several hours, emergency managers are working to resolve several critical issues, including a quick decision on the type of personal protective equipment and devices to be used to ensure the safety of the on-site responders (police, fire, medical personnel) and a decision as to evacuate or to shelter-in-place civilian populations in event impact areas. Over the next several to 12 hours, the emergency response team will be working to refine these evaluations and predictions, to assess the downwind impact zone in accordance with atmospheric transport and dispersion models so as to provide timely warning to threatened downwind populations, and to provide support for recovery efforts involving response personnel entering or re-entering affected areas. (Box 2.1.)

The time beyond roughly 12 hours following an event typically represents the transition period from crisis management to some degree of sustained managed response and the beginning of recovery activities. Of course, for long-lived chemical, biological, or nuclear releases, the response and recovery activities overlap significantly. As transport and dispersion models are supported by a more complete database of detailed atmo-

---

BOX 2.1
Secondary Users of Dispersion Information

In addition to the various types of first responders identified, there are a host of potential "secondary" users of information about atmospheric dispersion of hazardous agents. These may include public health officials, state and regional poison centers, hospitals, and non-governmental organizations that provide care and shelter for affected populations (Appendix D). There also may be numerous inquiries from the news media, political officials, and members of the legal and judicial communities (e.g., regarding Federal Bureau of Investigation forensic investigations). In most situations, it is best for those involved in atmospheric observational or modeling support to defer direct interactions with these secondary user groups to the established emergency response organizations.

---

[1] Dosage is the dose expressed as a function of time and the organism being dosed; for example, it can be expressed as milligrams per kilogram of body weight per day ($mg\ kg^{-1}\ day^{-1}$).

spheric observations and contaminant monitoring measurements, the response tasks also become correspondingly precise in terms of determining requirements for personal protective equipment, exposure[2] estimates across finer spatial resolutions, and conesquence assessment.

Especially during this period, the modeling support team must be familiar with non-technical aspects of the emergency management team's decision-making process. The decision makers not only need access to the best atmospheric transport predictions, but they also require reasonable estimates of the variability and confidence levels of results. They typically must reach some balance between safety concerns under a worst-case lethality scenario and the expense and other consequences accompanying over-reaction to such a scenario. In addition, while model output generally can be no better than data input, even the most sophisticated emergency response team members caution that models requiring input data from the end user that the user does not understand or cannot immediately provide will result in the model's being quickly discarded. Model providers must work diligently to assume the perspective of the end user by always asking, "What is needed, and how much is enough?" They also must recognize that the emergency responder often will have to reach a decision based upon whatever incomplete or imprecise information is available at the time. Transport modeling must be designed to provide the best support available even under the most difficult and limiting circumstances.

Finally, the emergency response team does not enjoy the luxury of a posteriori statistical analysis and comparison of differences accompanying competing atmospheric models. They need definitive support—without excessive complexity, caveat, or confusion—to directly address the decisions they must make, on the timetable on which they must make them. The burden of interfacing the atmospheric transport models to the decision-making needs of the emergency response team generally must fall upon the modeling community. A regular series of "tabletop," functional, and full-scale event simulation exercises (Box 2.2), bringing together emergency response teams and members of the atmospheric modeling and observational communities, would greatly benefit all parties involved and facilitate the development of a common set of data interface and decision support protocols.

The emergency responders who participated in the workshop uniformly agreed that in real emergency events, the atmospheric modeling community should speak with a single voice. There is general dissatisfaction with the large number of seemingly competitive atmospheric transport models and services now supported by various agencies. Conversely, there is wide agreement on the value of having a single point of contact (preferably reachable through a 1-800 phone number) that can provide a clearinghouse of information about the available observational and modeling support and immediately connect first responders and emergency managers to the appropriate centers of technical

---

[2] Exposure is the concentration, amount, or intensity of a particular agent that reaches the target population, usually expressed in numerical terms of substance concentration, duration, and frequency (for chemical agents and micro-organisms) or intensity (for physical agents such as radiation).

> **BOX 2.2**
> **Tabletop Exercises**
>
> Regular day-to-day interactions among the relevant players in an emergency response action (e.g., first responders, dispersion modelers, meteorologists) are necessary to ensure an effective working relationship. One particularly useful form of interaction is the tabletop exercise—a common method of training in the emergency management community, wherein participants plan and discuss responses to a given emergency scenario and sequence of events that may unfold during the course of that response. Tabletop exercises provide an opportunity to familiarize personnel with emergency response plans and to identify the roles and responsibilities of various individuals and organizations under those plans. Such exercises also provide a useful forum for allowing all of the different organizations involved in the response to a major incident to get to know each other and to work together. Tabletop exercises provide an effective means to educate personnel and practice emergency response without committing large amounts of time and resources. However, the need for at least occasional full-scale field exercises will remain, as they are essential for testing the appropriateness and execution of established procedures, the operability of models, and the interactions among all the various players involved in an emergency response action.

expertise. Of course, the usefulness of such a system will first require a comprehensive understanding of customer needs and the capabilities of various existing dispersion modeling centers. Such a resource may be especially valuable in smaller cities, towns, and rural areas, where first responders (who are often volunteer firefighters) may have little information about how to obtain immediate assistance.

Additionally, because of conflicting concerns over liability for decisions made and actions undertaken during the difficult first few hours following a terrorist attack, many in the responder community urge that atmospheric dispersion modeling and prediction be managed as a federal service.

As discussed in greater detail in Chapter 4, most models predict the average dispersion (over a large number of realizations of the given situation) and not the event-to-event variability about that average. As a result, even a good atmospheric transport model may have single-event errors of more than a factor of ten. In determining evacuation zones based upon estimates of lethality dosage, fluctuations of this magnitude represent substantial human health risks. It is important that atmospheric models applied to individual atmospheric releases provide predictions with clearly stated uncertainties.

There is an opportunity to improve the overall understanding of atmospheric transport and dispersion modeling by advancing research in this field and by synergistically combining the different techniques and approaches, as described later in this report. The subtleties of choosing among models, and determining how they are to work together under changing atmospheric conditions and output needs, must remain the challenge of a nationally coordinated effort and not be left as a responsibility of emergency response managers in the field. These end users have requested modeling outputs that offer simplicity, repeatability, scalability, and timeliness. With appropriate attention, the committee believes atmospheric transport and dispersion modeling can meet these needs and substantially enhance our national emergency response capability.

RECOVERY AND ANALYSIS

There is no specific timetable that establishes when recovery from a harmful atmospheric release begins. Because of the nature of transport, event recovery may be well underway in areas initially affected by the release as the hazardous agents reach new locations downwind. Atmospheric transport models should provide accurate prediction, warning, and exposure assessments for these later-time concerns (Box 2.3).

During the recovery period, health care workers will become much more active in reaching and caring for the injured. Atmospheric modeling predictions of exposure expectations will help the health care community assess the size of the needed response and the accumulation and allocation of necessary resources to deal with the events. Also, during the necessary triage of incapacitated and ambulatory injured, model predictions of exposure may influence the interpretation of symptoms and treatment modalities.

Emergency response workers will continue to monitor contaminant exposure levels and confirm when an area is safe to reenter, prescribing personal protective equipment as a function of exposure risk. Modeling efforts may prove especially valuable in highlighting possible geographic or structural areas capable of capturing and maintaining dangerous concentration levels of persistent hazardous agents. At some point, those who have been evacuated from their communities will be allowed to return home. The timing of such actions will depend in large part on the decontamination needs of the built environment, including likely contaminant collection sites such as storm drains, sanitary sewers, and building basements. In some cases, long-term environmental monitoring and restoration of natural lands, plants and animals, and waterways may become necessary.

BOX 2.3
Examples of Model Application to Post-event Analysis

Several examples discussed at the workshop illustrated how dispersion models can contribute to post-event exposure assessments over a wide range of spatial scales. For instance, following the Chernobyl nuclear accident, regional- and global-scale transport models were used to assess what locations and populations may have been exposed to radioactive fallout (Appendix G). Following the attacks on the World Trade Center, urban-scale dispersion models were used to assess what neighborhoods were exposed to the plumes of smoke emanating from the fires (Appendix F). Variable-scale dispersion modeling studies were carried out after the Persian Gulf War to predict dosage probability distributions (Appendix E).

KEY FINDINGS AND RECOMMENDATIONS

Atmospheric observations and dispersion models must interface seamlessly with the needs of emergency responders. Emergency response managers would benefit from training that conveys the strengths and weaknesses of existing observational and dispersion modeling tools and the situations under which various types of tools perform best. Conversely, dispersion modelers and meteorologists would benefit from learning how nowcasts and forecasts are used in emergency response situations. **"Tabletop" (i.e., roundtable discussion and planning) event simulation exercises should be convened regularly to bring together emergency response teams and members of the atmospheric modeling and observational communities to help establish and exercise a common set of data interface and decision support protocols.**

Emergency responders face a confusing array of seemingly competitive atmospheric transport model systems supported by various agencies, and in many cases, they do not have a clear understanding of where to turn for immediate assistance. **A single federal point of contact should be established (such as a 1-800 phone number) that could be used to connect emergency responders across the country to appropriate dispersion modeling centers for immediate assistance.**

Emergency managers need a realistic understanding of the bounds on the uncertainties of dispersion model predictions. Dispersion model predictions of the concentrations for a given release need to be accompanied by a prediction of the event-to-event variability in that situation. **Dispersion modelers should use ensemble modeling or other approaches that quantify not only the average downwind concentration distribution in a given situation, which is interpretable as the most likely outcome, but also the event-to-event variability to be expected. The specific formats of the information presented should be developed in close collaboration with users of this information.**

3

Observational Capabilities and Needs

In the event of C/B/N releases, observations will play a key role in tracking the agent and anticipating its spread, its dilution rate, and the projected exposure of the population. The unpredictability of such events dictates the need for instruments to be in place before a release occurs, mobile instruments that can be rapidly deployed to the site of the events, and longer-term deployable instruments to determine the total impact and indicate when affected areas are safe again.

It first is necessary to locate the plume and determine its composition, if possible. Identifying the composition will require knowledge of the "background" levels of agents of interest to prevent false alarms or misidentification of plume boundaries. Forecasting the plume direction requires knowledge of the wind field, which can be influenced strongly by local flows. Some measure of turbulence intensity is needed to determine how fast the plume will spread both vertically and horizontally. The amount of C/B/N agent that will settle to the surface is determined by the turbulence level and the composition of the plume in the case of dry deposition, and by the composition of the plume and scavenging by clouds and precipitation in the case of wet deposition. If the source of a release remains unknown after the events, downwind observations in combination with trajectory models are needed to back-calculate the source location (and possibly even the source strength). For instance, in the Chernobyl event, hemispheric-scale observations were used to detect the problems and identify its source. In the case of a terrorist attack, such observations also may play a valuable role in ongoing criminal investigations of the dispersal methods and individuals or groups involved.

The complementary role of observations and models will change through the course of a C/B/N release. Initial response will likely rely entirely on direct observations. Just after the event, forecasters (that is, those employing observations and modeling tools to make dispersion predictions) will rely on simple extrapolation of local data. Eventually, the forecaster will blend numerical models and nowcasting[1] techniques that draw on the relevant data. On time scales of hours to days or weeks, numerical models will be used to a much greater extent, although observational data from synoptic-scale

---

[1] Nowcasting refers to short-term weather forecasts, generally out to six hours or less.

networks will have to be assimilated continuously into the models. While the general improvement of synoptic networks and weather forecasts is important, the committee's emphasis is on special needs for tracking and predicting C/B/N plumes on a local scale in the first hours after release. The emphasis eventually may shift to regional and global models, but the observations required for model forecasts on these scales are beyond the scope of this report.

PLUME IDENTIFICATION

Identifying and following a C/B/N plume is critical both for real-time response efforts and for model nowcasting and forecasting. The first detection of a C/B/N agent could be from a directly observed plume, from fixed sensor instruments (Box 3.1), or from sickened humans or animals. Once a C/B/N event has been identified, there are a number of technologies that could be employed for three-dimensional sampling and tracking of a plume (Appendix C), although there are limitations associated with each. Scanning lidars, which use optical radiation much like radars use microwave radiation, can be used to sense aerosols and some trace gases out to 10–20 km (Plate 1) (Banta et al., 1999; Darby et al., 2002), although complex terrain and urban environments can limit the line of sight for such instruments. Furthermore, lidars are expensive and currently only available for civilian use in a research context. Microwave radars can identify and follow some plumes, depending on their composition.

There are a number of mobile platforms that can be used to deploy such instruments. For instance, boundary layer profilers or small radars can be readily mobilized on trucks or trailers. Helicopters can be used to lower tethered sensor pods into affected regions[2] as long as care is taken not to induce turbulence and considerably stir up the plume. Unmanned aerial vehicles (UAVs) can carry sensors that directly sample plume composition into regions not easily accessible to ground-based lidars or radars, such as urban canyons. To be effective however, these types of mobile platforms would have to be readily available for immediate deployment at the site of the emergency. For instance, if the UAV release point is far from the C/B/N impact region, the time required for deployment may limit the usefulness of this approach. In the absence of high clouds, satellites can track visible plumes. Gases have characteristic signatures in the infrared, so new satellites sensing the infrared spectrum at higher wavelength and spatial resolution (Mecikalski et al., 2002) will have increased capability for possibly identifying some plume constituents.

WIND—LOCAL FLOWS

Wind measurements are needed to track a near-surface C/B/N release in real time and to provide input for dispersion models, numerical weather prediction (NWP) models,

---

[2] Helicopters that operate routinely in urban areas for purposes such as traffic reporting have frequently been used as platforms for other types of observational equipment (e.g., such as that used in air pollution studies).

BOX 3.1
C/B/N Sensors

C/B/N sensors are an integral part of any system for tracking and predicting the dispersion of hazardous agents. There has recently been a concerted effort to accelerate the development and deployment of these sensors, and it is likely that C/B/N sensor data will become increasingly available. Assimilation of these data—both for model development and real-time operations—is a key new capability that will help optimize dispersion predictions. The National Academies' Board on Chemical Sciences and Technology recently held a workshop that included some discussion of C/B/N sensor technologies. Below is a summary of some points raised at that workshop (NRC, 2002a).

Because the amount of agent used in an attack can be relatively small, real-time sample collection, concentration, and analysis all are crucial issues for detection of C/B/N agents. There is a great deal of ongoing research to develop specific, sensitive, fast, and portable sensors. For example, new microfluidics technologies to accurately control the flow of liquids on a small (millimeter-scale) device have been key to the development of low-cost, portable packages used by first responders and emergency medical personnel to rapidly analyze small samples. To extend miniaturization to the sampling and concentrating of airborne particles, advances are needed in flow and handling of small volumes of gases.

Analytical techniques for the detection of some chemical and explosive agents are well established, including mass spectrometry and ion mobility spectrometry. However, current methods need to be improved and expanded to allow detection of many other potentially important chemical agents. Some promising technological developments include:

- fiber optic-based sensors that provide rapid response to a variety of chemicals at trace concentrations;
- flow injection analysis on a microelectromechanical system platform that provides high sensitivity and selectivity within hundreds of seconds from a small sample volume; and
- micromachined gas chromatography sensors that aid in real-time chemical sensing of toxic gases.

Some key challenges in this field include improving detector sensitivity and specificity and reducing the power drain so that smaller-size batteries can be used. Advances in microelectronics that have enabled the fabrication of compact, low-power devices and new miniaturization techniques, nanofabrication tools, and fundamental materials chemistry should allow significant advances to be made in the coming years.

There also have been a number of promising advances in sensors for biological agents, for example:

> - Developments in using mass spectrometric techniques to identify large biomolecules likely will prove important in identifying biological warfare agents.
> - Recent work has included concentrating and identifying bacterial pathogens such as anthrax spores based on protein biomarkers.
> - Components of biological systems, such as an antibody or a biomimetic membrane, have been incorporated into sensors for biological or chemical toxins.
>
> The most significant challenges in this area are to develop efficient approaches to collect, separate, concentrate, and process samples and to develop miniature devices that work under ambient conditions. General research into the biochemistry of agents and the rapid identification of agent pathogenicity is needed for developing the ability to respond to new threats such as artificially bioengineered agents.
>
> Knowledge of background levels of radioactivity, hazardous chemical, and spores and other bioagents is necessary to isolate real events from false alarms. Air pollution monitoring data provide information about the background concentrations of some hazardous chemicals, and the background radioactivity may be known in some areas with a history of mining or processing of radioactive materials. However, there generally is no ambient monitoring for most C/B/N agents of concern[3], and for many important hazardous agents, techniques for determining background levels do not even exist. It is particularly difficult to distinguish toxic biological agents from the harmless biological compounds ubiquitous in our environment or from naturally occurring toxic biological agents.
>
> Within the scientific research community, there is a general lack of knowledge about many of the characteristics of pathogenic or toxic agents. Innovations in this area could be encouraged by making available to researchers an extensive database on the properties of pathogens. A logical home for such a database might be the Chemical and Biological Information Analysis Center at Aberdeen Proving Ground, where extensive information about chemical and biological agents is maintained.

and hybrid (dispersion-NWP) models. Near-surface and low-level winds (i.e., surface and boundary layer winds during daytime; winds within the lowest few hundred meters at nighttime) are most critical in the first hours after a C/B/N release. Local terrain can cause strong spatial variations in wind speed and direction, creating local flows that are, in turn, disrupted by the presence of buildings. Local flows (e.g., mountain-valley winds, land and sea breezes, horizontal eddies caused by deflection of the wind by terrain) lead to large deviations from what would be expected for flat, uniform terrain. While some local flows can be easily observed, flows such as the Catalina Eddy in the Los Angeles area (Bosart, 1983) and the Denver Cyclone (Wilczak and Glendenning, 1988) were

---

[3] However, there was a recent announcement that air quality monitoring stations around the country will be augmented with sensors to monitor for anthrax, smallpox, and other deadly biological agents (Miller, 2003).

FIGURE 3.1 This 1973 photograph showing a "smoke run" at Brookhaven National Laboratory graphically illustrates the point that wind currents can move independently (and even in opposite directions) in separate layers of the lower atmosphere. The flatness of the smoke plumes suggests little turbulence, thus little vertical mixing in this case. (SOURCE: Courtesy of Brookhaven National Laboratory).

unknown until dense meteorological observations were taken. Such circulations are of the order of 100 km across or less—too small to be well observed by standard surface weather observing networks. Other local- to mesoscale (roughly 10–100 km) circulations can develop independent of the local terrain. Those associated with convective storms are readily identified and closely watched, but we are only dimly aware of other less apparent circulations that persist after a storm has died or that develop in response to subtle differences in land cover or soil moisture. Such circulations will need to be measured on a routine basis.

Local flows can carry C/B/N plumes in unexpected directions; for example, in some conditions a plume can stagnate or suddenly reverse direction. As illustrated in Figure 3.1, flows can vary dramatically depending upon the height above ground (Appendix C). Thus, forecasters must make an effort to understand the behavior of local flows in their areas. Such advance work enables optimization of measurements and trains the forecaster to interpret measurements and model output correctly. More needs to be known about how local flows develop in geographical areas with subtle variations in terrain or land use. More, too, needs to be known about how local terrain or urbanization affects frontal passages, mesoscale convective systems such as squall lines, and other weather phenomena (Appendix F). The same reasoning applies to the application of forecast models to local flows—good forecasters understand the strengths, weaknesses, and biases of the numerical weather forecast models they use on a daily basis. However, these models typically are applied to larger scales than those of interest in the context of a C/B/N release, and some forecasters have little experience with the behavior of the localized models needed to follow a C/B/N release during the first several hours. Forecasters in highly critical areas should monitor local flows routinely and compare this behavior to the prediction from mesoscale models to develop experience in forecasting local flows.

Tracking a C/B/N event in real time requires instrument arrays that can document the local flows determined by terrain or buildings. Low-level upstream winds may be useful for initiating numerical simulation of the airflow, particularly in areas with complex terrain. After several hours to days, depending on the weather and the extent of the release, winds at higher altitudes can become important in tracking and forecasting the plume dispersion to determine the potential risks faced by downwind populations. Mesoscale circulations can be readily detected by dense networks of surface weather observation towers that are typically instrumented to measure wind direction and speed, pressure, temperature, and humidity. To resolve a circulation, the station spacing has to be much smaller than the circulation scale—at least three observation sites across the circulation are required to detect it, and at least six are needed to resolve the circulation reasonably well along one direction. Circulations in "clear air" also are visible from Doppler radar through the presence of insects or strong humidity and temperature fluctuations, the latter of which make air visible to the radar by modifying the atmosphere's refractive index.

The U.S. meteorological observation network—including federal, state, local, private, and research networks—can meet a significant fraction of this daunting need. The Joint Office of Scientific Support at the University Corporation for Atmospheric Research maintains a database of available observational networks as part of its mission to support field programs (see http://www.joss.ucar.edu/gapp/networks). Also, the University of Utah has teamed with the National Weather Service (NWS), other government agencies, and the private sector to collect data from surface networks throughout the western United States for purposes of research, education, and operational support (see http://www.met.utah.edu/jhorel/html/mesonet/). There are dozens of tightly packed surface-tower arrays or single towers around the country run for monitoring air pollution or highway conditions, storm forecasting, research, K-12 education, or television weather programs. One such network, the Oklahoma Mesonet (Brock et al., 1995), is described in Box 3.2. Additionally, about a dozen universities operate buoys that provide environmental information offshore (Mesonet, 2002). With the increasing number of regions establishing mesonet systems, it would be useful to have one central focal point for coordinating the real-time acquisition and quality assurance of data from these networks. Furthermore, the development of "universal" software would allow easier access to and greater usability of these data.

> **BOX 3.2**
> **Example of a Multiuse Observational Network**
>
> The Oklahoma Mesonet (http://www.mesonet.ou.edu/) is a world-class observational network designed by scientists at the University of Oklahoma and Oklahoma State University. It includes 114 environmental monitoring stations distributed across the state. Each site includes a set of instruments, located on or near a 10-m tower, that measure parameters such as air temperature, relative humidity, wind speed and direction, barometric pressure, rainfall, and solar radiation. The observations are transmitted to a central facility every 15 minutes, 24 hours per day, year-round. The Oklahoma Climatological Survey receives the observations, verifies the quality of the data, and provides the data to Mesonet customers. It only takes 10 to 20 minutes from the time the measurements are acquired until they become available.
>
> The Mesonet data are applied for a wide array of uses, including weather forecasting, education and scientific research, and planning for agriculture, energy supply, and transportation. In addition, the network already is being used by public safety agencies for tracking hazardous material release incidents, as described in Morris et al., (2001) and as highlighted in the following quote: "Mesonet is without a doubt among the most important data sets we use at the National Weather Service Forecast Office. In addition to routine forecast and warning operations, the Mesonet is invaluable for handling various disaster support situations including wildfires, chemical spills, and catastrophes like the Oklahoma City Murrah Building bombing" (David Andra, NWS Forecast Office, Norman, Oklahoma).

The NWS, the Federal Aviation Administration, and the military operate a national network of Weather Surveillance Radar-1988 Doppler (WSR-88D) radars (see NRC, 1995; Figure C.1), and radars operated by television stations may be used to provide additional coverage with the permission of the stations. For low-level scans, radar clear-air wind field coverage at the surface is limited by Earth's curvature to a maximum distance of about 50 km (Plate 2). The current pre-programmed radar scans might not be optimum for determining the low-level wind field with the detail needed to follow a C/B/N release. However, one of the possible enhancements to the radar network (NRC, 2002b) is to supplement the current WSR-88D network with subnetworks of smaller, less powerful, and less expensive short-wavelength (3- and 5-cm) radars to provide more low-level coverage. New scan designs for C/B/N responses also should be considered.

The National Oceanic and Atmospheric Administration (NOAA) operates a network of vertically pointing 400-MHz band radar wind profilers across the central part of the United States (Martner et al., 1993), which provides winds at heights from 500 m to about 16 km, but the lack of data at lower levels and the relatively coarse vertical resolution limit the usefulness of these profilers for near-surface applications. There are, however, several 900-MHz band boundary layer radar wind profiler networks being used for research in Oklahoma, Kansas, Texas, California, and elsewhere. Some are combined with Doppler sodars to obtain winds down to about 30 m off the surface. Radar wind profilers with radio acoustic sounding systems (RASS) provide estimates of the temperature profiles through slightly shallower depths than where winds are measured. Aircraft winds are available from the surface to jet cruising altitude through the Meteorological Data Collection and Reporting System (MDCRS; Moninger et al., 2003).

Once a C/B/N release occurs, nearby wind sensors (as well as simple "intuitive" indicators such as flags and the trajectories of visible smoke plumes) will be used to help locate the site and spread of the release. Data from the fixed observational arrays discussed above will be useful, but additional observing systems may need to be mobilized to cover some areas affected by a C/B/N release. If the release is in a city with tall buildings, estimates of the wind in urban canyons will be urgently needed, since model winds will likely be woefully inadequate. Video surveillance cameras, or "web-cams," could provide information on the motion of visible plumes. Inexpensive optical cross-wind sensors could be used to sense winds in selected urban canyons. These instruments measure the average wind component transverse to their optical axis usually to a distance of 200 m to 1 km. They are ideally suited to be deployed between buildings, and they give the down-street flow, possibly at multiple stories to estimate vertical wind structure. By slanting the path along the street, estimates of the average wind vector can be determined.

Mobile sensors could include scanning Doppler lidars or radars and UAVs. These platforms complement one another, since lidars can "see" aerosols in relatively clear air, even if there are not enough scatterers for the radars to detect a signal. Although lidars may not be useful in the presence of clouds and rain, radars can derive winds from insect scatterers, precipitation, and possibly the plume itself. Radar and lidar are limited to line-of-sight data, but in some contexts, UAVs may be able to fill in wind field observations between buildings. Mobile Doppler wind profilers could provide data between 150 m and 1–2 km above the ground. Depending on the instrument type and application, mobile sensors could be carried on trucks, aircraft, helicopters, or boats. These mobile instruments will be needed most urgently in the first minutes to hours after a release, hence, rapid access is critical. Such instruments must be located close to threatened areas and be available for immediate deployment, which may be feasible only in a few select locations.

DEPTH AND INTENSITY OF TURBULENT LAYERS

The depth of the turbulent layer near the ground and the intensity of the turbulence (and hence mixing) also are in the minimum data set necessary to estimate the transport and dispersion of a C/B/N release, to determine how the plume will spread and mix vertically as well as horizontally (i.e., three-dimensionally) (Appendix C).

The heights and depths of turbulent layers are shown clearly in reflectivity profiles from 900-MHz band radar wind profilers at altitude ranges from 150 meters to a few kilometers (Plate 3). If collocated with radar wind profilers, the RASS provides temperature profiles with the same height restrictions, although its use may be compromised by its noisiness. Sodars can provide information similar to that of radar wind profilers near the surface. Doppler lidars can detect layers of turbulent mixing vertically up to 10–15 km (Plate 4a) as well as horizontally to a distance of 10–20 km (Plate 4b and Plate 5) (Rothermal et al., 1998; Banta et al., 1999; Darby et al., 2002). If there is an airport nearby, MDCRS temperature profiles from commercial air carriers can be used to identify layers where mixing is likely.

For a C/B/N release occurring a few hundred meters above the surface (for instance, a release from a crop duster), the urgent question is how quickly the hazardous material will mix or settle to the surface. During the daytime, there often is strong turbulence throughout the lowest few hundred meters, which would quickly mix the material to the ground. However, when there is warm air moving over a cold surface or when cloudy skies prevent strong heating of the ground, there may be little mixing. The stable warm-over-cold air temperature layering that often occurs at night can isolate the surface from such releases, but breaking waves or surges of cold air can initiate mixing between the turbulent layer and the surface, or the wind change between the layers can increase enough to promote mixing (Box 3.3). Surges of cool air are visible in surface-station arrays, and breaking waves are observable using scanning lidars. Temperature and wind profiles can be determined from MDCRS data, UAV data, RASS data, or special radiosonde soundings. As illustrated by Figure 3.1, vertical profile information is critical, since a release at the surface may have a vastly different outcome than a release occurring at various heights above the surface.

DEPOSITION AND DEGRADATION

A C/B/N release must be viewed as more than just an atmospheric hazard, since the hazardous agents eventually will be deposited from the atmosphere onto surfaces such as buildings, soil and vegetation, and aquatic systems. Deposition patterns and the resulting impacts will depend heavily upon the contaminants' atmospheric residence time (which could vary from minutes to weeks, depending on particle size and other physical properties of the agent) and environmental viability (that is, how rapidly the agent's potency diminishes after exposure to ambient conditions). The deposition process occurs through one of several possible mechanisms including dry deposition, wet deposition (rain or fog), or gas-phase reactions with various surfaces.

Dry deposition of a hazardous agent shortly after release time can be estimated from plume location, concentration, and the turbulence level. Such agents may pose a "secondary hazard" if they are returned to the atmosphere or water supply by wind, rain, or fires and, in some cases, persistent agents may eventually propagate through ecological systems and the food chain. Appropriate sensors can be deployed to track these latent sources of potential harm, although tracking residual agents sequestered in isolated areas or adsorbed on various materials may involve a challenging assay process.

Wet deposition of a hazardous agent occurs when precipitation containing the agent falls to the surface (vertical deposition) or when cloud droplets containing the agent intercept the surface or vegetation on hill or mountain slopes (horizontal deposition). Moisture also plays a key role in the degradation of some hazardous agents; for example, degradation of nerve and mustard agents occurs via hydrolysis by aqueous aerosols and

> **BOX 3.3**
> **Daytime and Nighttime Mixing Patterns**
>
> When a toxic release occurs, mixing and dispersion patterns will differ significantly depending on such factors as the time of day the release occurs and the weather conditions at the time. Two extreme examples illustrate these effects.
>
> During a summer day with abundant sunshine, the sun heats the ground and, indirectly, heats the air close to the ground. This warmed air rises, cooling and entraining cooler air as it does so (e.g., Plate 4a). The buoyant plumes will continue to rise and cool until they are cooler than the air around them, at which point further upward motion is cut off. Air from the surrounding area flows in to replace the buoyant plumes, and this leads to efficient vertical mixing in the lower atmosphere. The layer at which this turbulent mixing occurs is called the atmospheric boundary layer. In this type of daytime scenario, cloud cover tends to retard heating and acts to reduce thermal mixing.
>
> During a clear winter night, the ground radiates heat energy to the surrounding atmosphere (e.g., Plate 4b). The air near the ground becomes cold relative to the air above it, which leaves the coldest air "trapped" near the ground. Turbulence and mixing are suppressed by this thermal stratification. Under these conditions, the atmospheric boundary layer commonly is less then 100 m deep and can occasionally be as shallow as a few meters. Sometimes the air is more turbulent a few hundred meters above the ground than at the surface (e.g., when the wind blows over the top of a stable lower layer); thus, important mixing can take place above the boundary layer. In a nighttime situation, cloud cover tends to suppress radiative cooling and thus mitigate the cold air trapping near the surface. It is important, then, to monitor not only the height of the boundary layer, but also the temperature and wind throughout the lower few hundred meters.

damp surfaces. Thus, it becomes important to monitor clouds, precipitation, dew points, and soil moisture along the track of the C/B/N plume. Clouds are routinely monitored from meteorological, environmental, and other satellites. The Doppler radar network does an excellent job of documenting precipitating storms over the continental United States, but does only a fair job of estimating precipitation amount. In the future, precipitation data from the Global Precipitation Mission satellites (Shepherd and Mehta, 2002; Shepherd and Smith, 2002) and stream-gauge data will aid in estimating precipitation. Progress also is being made in running mesoscale models using assimilated radar data, providing another avenue for future improved estimates. Satellite views of the low clouds associated with horizontal deposition may be obscured by higher clouds, so forecasters and emergency managers will have to rely on models and wind field observations to project where the plume will intercept hills or mountains. Following deposition, hydrological and ecological models and observations may be required to provide information about subsequent dispersal of the hazardous agents through the environment.

KEY FINDINGS AND RECOMMENDATIONS

The most basic observations required for tracking and predicting the dispersion of a hazardous agent include identification of the plume; characterization of low-level winds (to follow the plume trajectory); characterization of the depth and intensity of the turbulent layers through which the plume moves (to estimate plume spread); and identification of areas of potential agent degradation and dry or wet deposition. Table 3.1 summarizes the observations and instruments most useful for a response to a C/B/N release.

The current array of surface observational systems needs to be better used and enhanced. Many surface stations are poorly exposed and have limited instrument quality control, and instrument locations are not necessarily optimal for model initialization or identification of local flows. Furthermore, it often is difficult to obtain the data from multiple observational arrays, especially in real time. **A comprehensive survey of the capabilities and limitations of existing observational networks should be conducted, followed by action to improve these networks and access to them, especially around more vulnerable areas.**

Doppler radar systems can be useful for estimating boundary layer winds, monitoring precipitation, and tracking some C/B/N plumes. The National Research Council (2002b) recommended evaluating the potential for supplementing current Doppler radar network with subnetworks of short-range, short-wavelength radars. This would enable better estimates and coverage of low-level winds, increase the likelihood of detecting C/B/N plumes, and improve precipitation (and hence wet deposition) estimates. **The committee supports this recommendation and further recommends that the design and data collection strategy of this radar network be optimized to include providing information for supporting response to a C/B/N release.**

Radar wind and radio acoustic sounding system profilers, which measure variations of the horizontal wind and temperature, respectively, with height and enable identification of turbulent layers, provide important information for response to C/B/N attacks and are relatively inexpensive and easy to maintain. **Wind and temperature profilers should become an integral part of regional and local fixed-observational networks.**

Mobile observational platforms can provide valuable information and fulfill multiple needs in the first minutes to hours after a hazardous release. Unmanned aerial vehicles can be used to measure wind and temperature profiles and to characterize turbulence where other platforms cannot easily reach. Mobile lidars and radars can, in some contexts, be used for plume tracking and wind field characterization. However, civilian instruments currently are available only for research use. **There should be continued development of portable scanning lidars and radars on airborne and**

TABLE 3.1 Observations and instruments useful for response to a C/B/N release. Details on some of the systems appear in Appendix C.

Observations	Reason	Instruments	Coverage	Needed Enhancements
Plume location	Determine or project affected population	Scanning lidar	*Vert:* 0 to 1-2 km *Horiz:* up to 10-20 km	Affordable, eye-safe
		Scanning radar (clear air)	*Vert:* 0 to 1-2 km *Horiz:* 10-50 km, depend on plume	More radars to increase low-level coverage (as proposed in NRC, 2002b); special scans and data processing to obtain low-level wind field
		Satellite visible or IR	Visible daytime only	
		UAV	Where sent	Available quickly to critical locations
Plume composition	Estimate exposure	Scanning lidar (fixed or mobile)	*Vert:* 0 to 1-2 km *Horiz:* up to 10-20 km	Affordable, eye-safe
		Satellite IR	N/A	High spatial and wavelength resolution. Several such satellites are planned, including the GOES[a]-Advanced Baseline Imager and Advanced Baseline Sounder (planned launch 2012), and polar-orbiting sites (planned launch 2007-2008), which will have higher horizontal resolution. (Mecikalski et al. 2002)
		In situ sensing from UAV or sensor pod attached to helicopter	Where sent	Available quickly to critical locations
Low-level winds	Document horizontal transport by local flows; model input	Multiple surface meteorological-tower arrays	2-10 m	*Present arrays:* useful arrays identified, improved exposure and quality control, reliable data transmission to users *Additional instruments:* add or move present stations for detecting local flows in critical areas
		Scanning lidars	*Vert:* 0.1-1 or 2 km *Horiz:* ~0.1km to 10-50 km	*Present arrays:* flexibility to do needed scans to follow C/B/N plume *Future:* greater low-level coverage through use of shorter-wavelength, low-power scanning Doppler radars, as proposed in NRC (2002b)
		900-MHz-band radar wind profilers	150 m to 3-5 km, 60-75 m vertical resolution	*Present arrays:* increase vertical resolution through better signal processing *Future:* add more radar wind profilers, supplement with Doppler sodars *Longer term:* replace with less noisy equivalent
		Doppler sodar	30-200 m at 5-m vertical resolution	Required loud sound pulses make them difficult to deploy; develop radar wind profiler with lower-altitude capability to replace sodar
		MDCRS soundings	Surface-12 km near airports	Availability to C/B/N event forecasters
		Scanning Doppler lidars	*Vert:* 0 to 1-2 km *Horiz:* 10-20 km	Affordable, eye-safe

	Video camera, web camera images of visible plume, flags, etc.	Near surface	Assess current capability (location, visible field, resolution) and then improve as necessary (e.g., install at locations with good visibility)
	Optical crosswind sensors for along street winds (scintillometers)	Across urban canyons	Install in critical areas; need development of a simple inexpensive version of existing systems
Fill in wind in critical areas	UAVs, mobile scanning Doppler radars, and Doppler lidars	Where sent	Available quickly to critical locations
Depth of turbulent layer(s)	900-MHz-band radar wind profilers	150 m to 3-5 km	*Present arrays*: increase vertical resolution through better signal processing. *Future*: add more units
Identify layer(s) through which plume will mix	Sodars	30-200 m	Replace with smaller radar wind profiler
Potential for airborne plume to mix to surface	Scanning lidars	0.1-2 km	Affordable, eye-safe
Identify mixing events propagating into area	Surface tower arrays	2-10 m	As for wind measurements
	MDCRS soundings	Surface-12 km	Available to C/B/N forecasters
Identify potential for mixing event	Special radiosondes	Surface-30 km	
	RASS	150 m-3 to 5 km	
Winds 500 m and above	400-MHz-band radar wind profilers	Surface-16 km at 300 to 900-m resolution	
Document horizontal transport; model input	Radiosondes	Surface-12 km	Provision for special radiosonde releases as needed
	Satellite	Where tracers are	Plumes and clouds can be tracked in visible or IR to provide winds; improve satellite wind-tracking capability
Dry deposition	Data or models used to estimate plume location		Estimate human, environmental exposure, plume depletion
Wet deposition	Radar, rain gauge, satellites, stream gauges		Merging datasets to get best estimate of rainfall; increased radar coverage; assimilating data into model

[a] GOES stands for Geostationary Operational Environmental Satellite.

surface-mobile platforms for research, and plans should be developed to make such instruments rapidly available for effective, timely use in vulnerable areas.

Local topography and the built environment lead to local wind patterns that can carry contaminants in unexpected directions. Observational networks must represent these local flows as faithfully as possible. Improvements in these networks can be achieved through routine data monitoring and comparison of observed flows with local- to regional-scale model simulations and through numerical modeling, including observing system simulation experiments. Studies should be performed over a range of weather situations and for both day and nighttime conditions. Such exercises will educate meteorologists about local flows and model capabilities; the resulting knowledge of what to believe when observational data and models convey different messages is vital in response to an emergency situation. **Efforts should be made to systematically characterize local-scale windflow patterns (over the full diurnal cycle) in areas deemed to be potential terrorist targets with the goals of optimizing fixed observations and educating those involved in developing dispersion forecasts about local flows and model strengths and weaknesses.**

Focused field exercises are needed to understand the behavior of modeled transport and dispersion in different weather regimes and C/B/N release scenarios, particularly for nocturnal conditions. It is not practical to verify dispersion and transport models for every area with comprehensive field programs, but for an appropriate range of meteorological conditions, physical modeling in a wind tunnel could assist in dispersion model evaluation and threat assessment. In addition, field programs conducted for other purposes, such as improvement of weather forecasting or understanding boundary layer turbulence, also can be useful. **There should be continued field programs focused on C/B/N release issues, and datasets from field programs with a C/B/N or related focus should be made available for testing and development of dispersion and mesoscale transport models.**

Some of the actions recommended above (i.e., enhancing fixed observing arrays, optimizing placement of surface stations and wind profilers, developing and deploying portable scanning lidars, UAVs, and radars) will be costly. **There should be prioritization of such actions based on identifying areas with the greatest need (e.g., highest population concentration, most complex flow, greatest likelihood for a terrorist attack, most vulnerable facilities). Every effort should be made to utilize such instrumentation for other (hazardous and non-hazardous) applications (e.g., to enhance air pollution monitoring, optimize agricultural practices, aid in severe-storm forecasting and highway network safety), thus sharing the costs and ensuring that the array will be continuously used, maintained, and quality controlled.**

COLOR PLATES

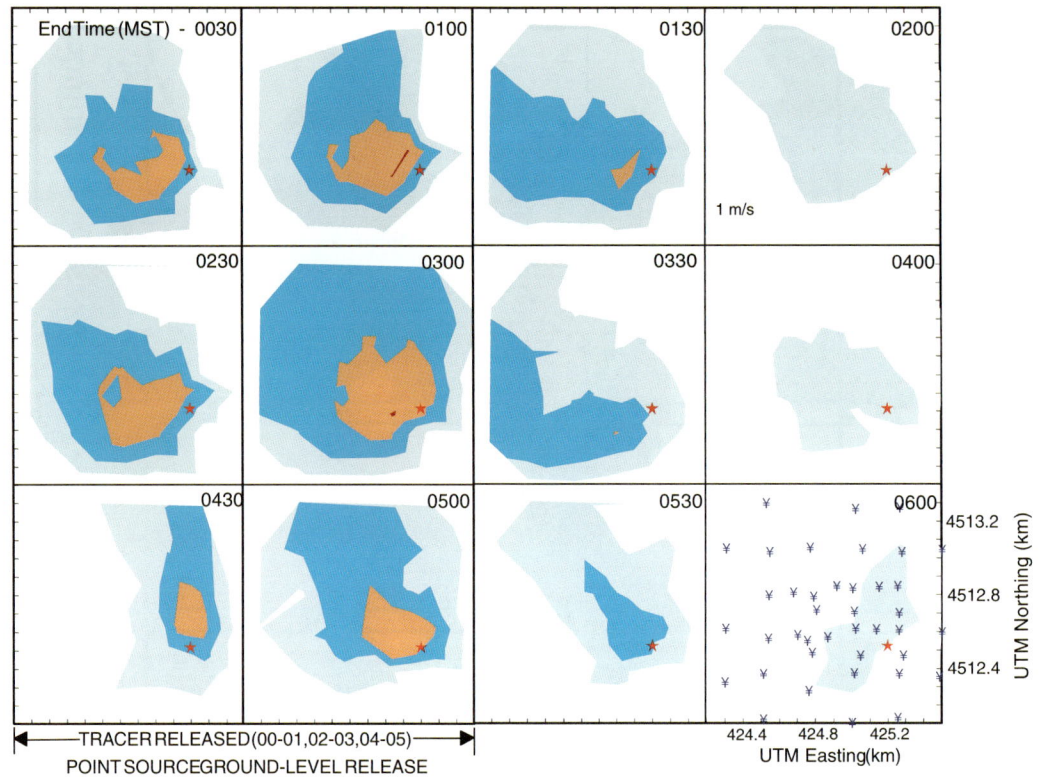

PLATE 8 Half-hour-average ground-level SF$_6$ plume concentration measurements from intensive operation period (IOP) 10 of the URBAN 2000 field campaign in Salt Lake City, UT. Data are taken during the early morning of October 26, 2000.

4

Dispersion Modeling: Application to C/B/N Releases

The dispersion of an effluent plume in the atmosphere is the result of transport by the wind field and distortion and mixing by turbulence. Figure 4.1a, a snapshot of a plume downwind of a continuous point source in a turbulent flow illustrates these turbulence effects.

Before the availability of modern computers, treatments of atmospheric dispersion focused on a time-average plume that varies smoothly in space, as illustrated in Figure 4.1b. In a flow where the time-averaged velocity and the turbulence properties are spatially uniform, this plume has a Gaussian concentration profile, and the downwind evolution of the plume width is related to statistical parameters of the turbulence. These concepts are the basis of the Gaussian-plume models[1] that have long been used to predict dispersion from continuous point sources in air quality applications. Today's computational fluid flow models can use a numerical grid with several hundred points in each of the three coordinate directions, and numerical techniques allow tailored grids that are finer near the source and coarser farther downwind. Such computational advances have led to a proliferation of the number and types of atmospheric dispersion models.

---

[1] Gaussian-plume models assume that the concentration of the agent downwind of the source (averaged over a large number of realizations of the given dispersion problem) has the form of the Gaussian, or "normal", probability distribution in the vertical and lateral directions. The amplitude and width of this "bell curve" are determined analytically by the rate of emission, mean wind speed and direction, atmospheric stability, release height, and distance from the release. Such models assume continuous and constant emission of agent, and they also generally assume flat terrain, no chemical reactions or absorption, and constant mean wind speed and direction with time and height.

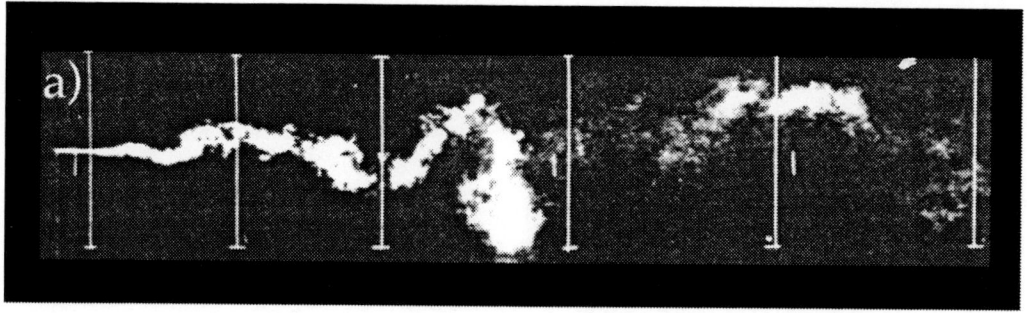

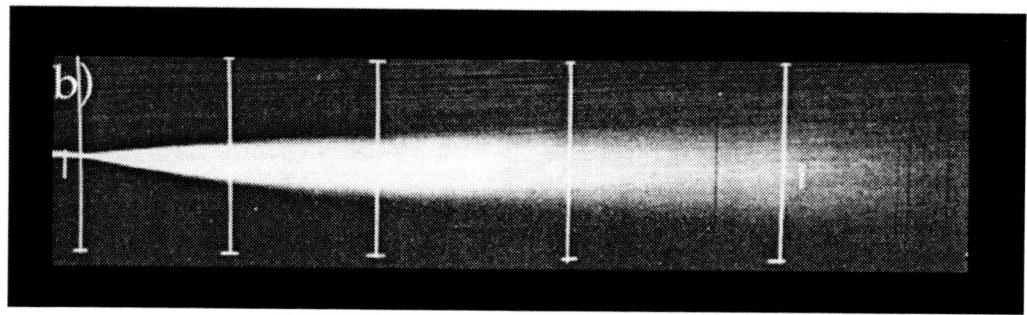

FIGURE 4.1 (a) A snapshot of the instantaneous plume downwind of a continuous source in a turbulent flow (horizontal cross section). (b) The time-averaged plume. (SOURCE: EPA Fluid Modeling Facility).

CATEGORIES OF DISPERSION MODELS

Atmospheric dispersion models can be broadly placed into categories using three distinguishing characteristics: (1) their *coordinate systems*, (2) their *wind fields*, and (3) the *type of averaging* used in developing the models from the underlying conservation equations.

Two *coordinate systems*, Eulerian and Lagrangian, are used. In a Eulerian system, the flow variables depend on time and on position in Earth-based coordinates. The Lagrangian system follows individual "fluid parcels" whose locations depend only on time. The Eulerian system is used in the vast majority of today's numerical flow models, including weather forecasting and climate models, but the Lagrangian system is naturally suited to dispersion problems.

The *wind field* in a dispersion model is, in some cases, defined by only a single value of the average wind at a specified height, such as in the simplest Gaussian-plume models. A step toward higher resolution is the incorporation of time-varying winds measured at several points within the domain. The highest-resolution dispersion models use a three-dimensional grid of winds calculated from a meteorological model.

The *averaging* used in the model development process is required for fluid-mechanical reasons. The equations governing flow and dispersion in the lower atmosphere have turbulent solutions with a range of spatial and temporal scales far wider than can be resolved on today's computers. Before the equations can be solved numerically in dispersion models, it is necessary that the range of resolved eddies be limited. This is done by ensemble or spatial averaging of the equations on which the models are based (as explained below). Of all the sorting criteria for dispersion models, the type of averaging applied to these governing equations has the broadest and deepest implications.

Ensemble averaging transforms the equations from a set describing a single episode of a turbulent dispersion problem to one describing the average of a large number of episodes (formally called realizations) of the problem. Figure 4.1a is a snapshot of a single realization. The bottom panel shows the time-average plume, which in this case is identical to the average of a large ensemble of such snapshots. The difference between the single-realization and ensemble-average effluent concentration fields is profound.

Gaussian-plume models for continuous releases are the oldest and simplest examples of ensemble-average dispersion models. They require a minimum of input information (average wind speed and direction, plus rudimentary information on whether the wind and temperature conditions favor turbulence and hence mixing, which allows diagnosis of the downstream growth of the Gaussian plume). There also are Gaussian models for finite-duration releases (called instantaneous releases) that can use an ensemble-average wind field derived from observations or computed through the dynamical equations.

In contrast to ensemble averaging, spatial averaging has quite a different effect; it produces an equation set describing a coarser-grained version of a realization of the problem. The solution fields retain their turbulent character at scales larger than that of the spatial averaging. One could visualize a snapshot of such a solution by removing the finer-scale detail from Figure 4.1a. Examples of spatial-average models include coarse-mesh meteorological models having a horizontal grid scale of tens of kilometers. Finer-mesh examples include mesoscale models with grid size on the order of 1–10 km, and large-eddy simulation (LES)[2] codes with grid size on the order of 100 m or less.

INTERPRETING AND EVALUATING DISPERSION MODEL OUTPUTS

In turbulent flow, the effects of slightly different initial conditions grow with time. As a result, two flows with nominally the same initial conditions eventually become quite different. This dependence on initial conditions has been found to be so sensitive that the initial conditions of a specific realization of a turbulent flow are unlikely to be known well enough to allow its reliable prediction. Thus, in the turbulent dispersion of effluent from a source, the downwind concentration patterns in two realizations of a given event will differ, the variation being more pronounced farther downwind of the source. For this reason, the output of a spatial-average dispersion model is properly interpreted not as a

---

[2] Large-eddy simulation is the term used for the numerical calculation of three-dimensional, time-dependent turbulent flows using spatial resolution sufficient to resolve the largest turbulent eddies.

prediction of the dispersion under the specified conditions, but rather as one of a range of possible outcomes under those conditions.

The timely availability of fine-scale wind-field measurements (i.e., with spatial resolution finer than a dispersing plume's local crosswind dimension and temporal resolution finer than the scale of its local time changes) could change this situation. Such data used in a spatial-average dispersion model could substantially reduce the realization-to-realization variability that now accompanies the prediction of the atmospheric dispersion of a short-term release. Currently, such measurements are not feasible except in special circumstances. Radar and lidar have high potential for enabling such applications in the future, although the time required to collect and assimilate high-resolution wind field data may continue to limit applicability to immediate emergency response needs.

As Figure 4.1 suggests, concentrations at any one point in a given realization can differ substantially from the ensemble average at that point. This further suggests that neither an individual realization nor the ensemble average of realizations is sufficient in general for assessing the detailed, short-term dispersion characteristics of hazardous materials. Both a prediction of the ensemble-average field (interpretable as the most likely outcome) and a measure of the realization-to-realization variations about this average field are needed. Some models (e.g., the Second-order Closure Integrated Puff, or SCIPUFF, model) predict the ensemble-average dispersion plus a measure of the variability of the concentration field from realization to realization (such as the variance or the probability density function[3]) (Appendix E).

Spatial-average models also allow probabilistic concentration forecasts. For instance, it is possible to vary the initial and boundary conditions and subgrid-scale physics of the dispersion model in order to generate an ensemble of forecasts of a given dispersion problem. This allows estimates of the spatially smoothed ensemble-average dosage field resulting from an instantaneous release as well as estimates of the probability that the dosage for any area will exceed given thresholds.

The concept of ensemble averaging need not focus only on the uncertainties in the turbulent dispersion process itself. One possible strategy for obtaining "end-to-end" uncertainties in a dispersion forecast is to create an ensemble average (and associated confidence levels) that includes a defined range of source and wind input variation by running multiple independent LES or physical simulations. This "brute force" approach cannot be applied directly to a real-time prediction, but it can be used to estimate uncertainties for a wide range of potential scenarios, and such a scenario database could provide an immediate first prediction for emergency responders. The scenario possibilities then could be updated with real-time model results as event-specific source and wind information become available.

---

[3] The probability density function gives the probability of occurrence of values of the function. In mathematical terms, the probability that a random function (C) lies in an interval (ΔC) around C_0 is $\beta(C_0) \Delta C$, and the integral from $-\infty$ to C_0 is the probability distribution of C.

Because of the long use of ensemble-average dispersion models in the air quality community, their evaluation techniques (Weil et al., 1992) are more advanced than those for spatial-average models. Time-average concentrations or dosages measured at individual points downwind of a source typically differ substantially from the predictions of ensemble-average models. However, it has also been found that concentrations averaged over one hour, say, retain a good deal of random variability (Figure 4.2). The interpretation is that downwind of a continuous source in the lower atmosphere, the time required for the convergence of a time-average concentration to the ensemble average can be much more than one hour. If so, one-hour-average observations would scatter substantially around predictions of even a perfect ensemble-average model, but the models are not perfect, and model physics errors also contribute to the observed differences. It can be difficult to apportion these differences between errors in model physics and the inherent statistical scatter, or "inherent uncertainty" as it is called in the dispersion-modeling community. Improved models have been found to have decreased scatter, as illustrated in Figure 4.3, and it is now evident that much of the scatter between predictions of the CRSTER, a standard Gaussian-plume model, and observations at the Kincaid site (Figure 4.2) was due to inadequate model physics.

The comparisons of model predictions and observations in Figures 4.2 and 4.3 are "paired in time and space," meaning that the observation and prediction associated with a given data point are for the same position in space and the same time period. Such comparisons typically lead to very large scatter, even with models having improved physics. For that reason it is common today to use quantile-quantile (Q-Q) comparisons instead. These are made by ordering the entire set of predictions by magnitude (from highest to lowest, say) and ordering the corresponding observations in the same way. Then the ordered predictions and observations are paired, the first with the first, the second with the second, and so forth, and the new pairs are plotted. Because of the ordering process, the observation and prediction associated with a given data point in general now do not correspond to the same position in space or the same time period. For this reason Q-Q comparisons are referred to as "unpaired in time and space." Figure 4.4 shows a Q-Q plot for the model at the Kincaid site. Its space-time unpairing greatly reduces its scatter from that in the conventional plot (Figure 4.3). This decoupling of the predicted and observed points can mislead the reader into thinking the model performs better than it actually does. What this technique does show is the ability of the model to predict the probability distribution of the time-averaged concentrations downwind of a continuous release.

Some unpairing of points also is done in testing other types of models. Figure 4.5 shows a scatter plot of observations downwind of a three-hour point release of sulfur hexaflouride (SF_6) versus the predictions of the VLSTRACK (Vapor, Liquid, and Solid Tracking) model. The observations are of the maximum dosage along the sampler lines 5–20 km downwind for each run, and the predictions are of the maximum dosage along the sampler lines for that run. In general, the predicted and observed maxima occur at different points on the line. If a dispersion model used in an air quality application yields a 1:1 line on a Q-Q plot, the probability distribution of its predictions over a downwind region agrees with that of the observations in that region. Often, more spatial specificity is not needed, and in such cases the Q-Q plot can be an effective model evaluation tool.

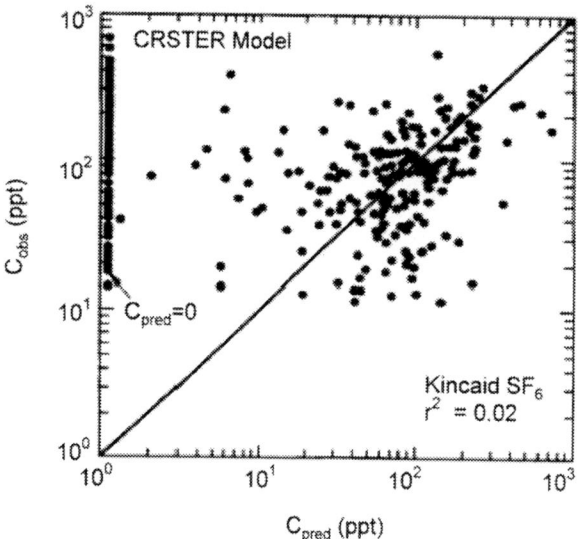

FIGURE 4.2 Observed versus predicted ground-level sulfur hexaflouride (SF_6) concentrations for the CRSTER Gaussian plume model at the Kincaid power plant. The observations are one-hour averages. The diagonal line corresponds to $C_{obs} = C_{pred}$. SOURCE: From Weil et al. (1997). Reprinted with permission from the American Meteorological Society.

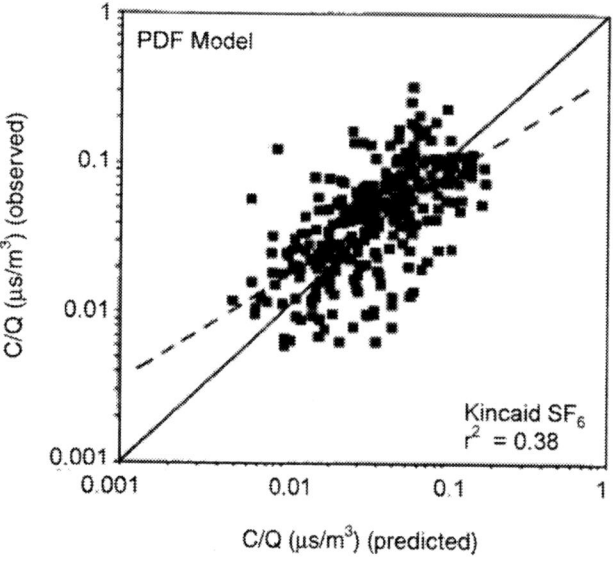

FIGURE 4.3 Observed versus predicted ground-level sulfur hexaflouride (SF_6) concentrations, normalized with the emission rate, for the PDF model at the Kincaid power plant. PDF is an ensemble-average model with improved physics. The observations are one-hour averages. SOURCE: From Weil et al. (1997). Reprinted with permission from the American Meteorological Society.

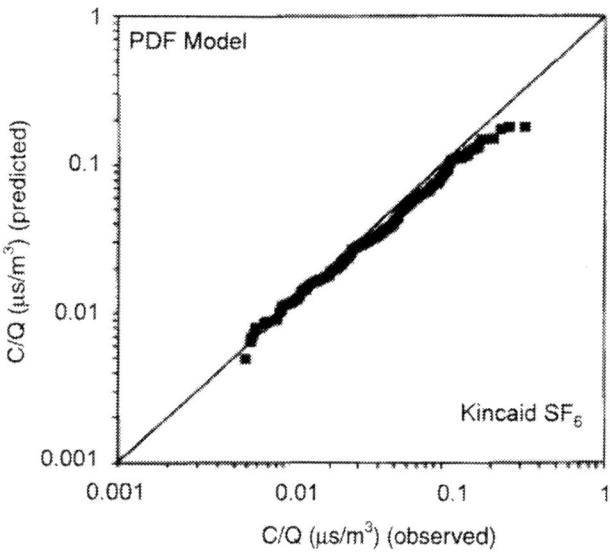

FIGURE 4.4 A Q-Q plot of the data in Figure 4.3. SOURCE: From Weil et al. (1997). Reprinted with permission from the American Meteorological Society.

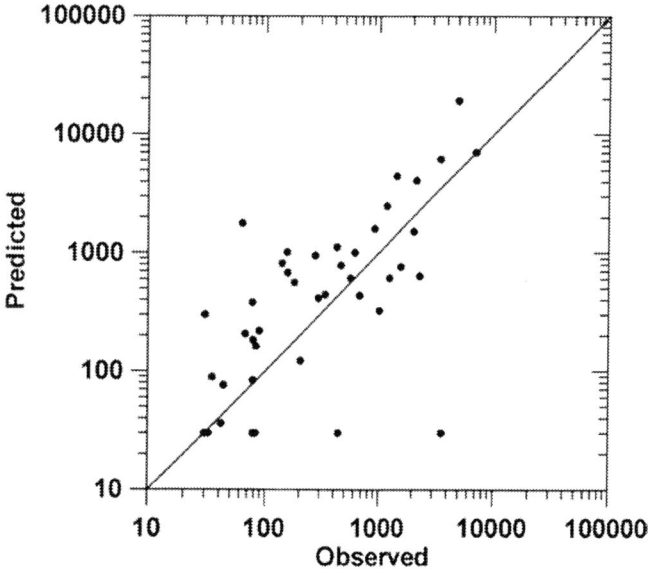

FIGURE 4.5 A dosage scatter plot for three-hour releases of sulfur hexaflouride (SF_6). SOURCE: From Chang et al. (2003). Reprinted with permission from the American Meteorological Society.

It is not clear that this is the case for the episodic models needed in emergency response applications, however.

It appears that a more useful test of episodic dispersion models would involve the probability density function (PDF) of the short-term-average concentration or dosage. This can be calculated from a sufficiently large set of observations at field or laboratory scale and is also provided by some advanced ensemble-average models (Weil et al., 1992). Ideally, any prediction of an episodic concentration field from a dispersion model should be accompanied by a prediction of its episode-to-episode variability quantified by the variance or, better yet, its PDF.

OVERVIEW OF C/B/N DISPERSION MODELING SYSTEMS

The workshop discussions and presentations addressed many facets of how transport and dispersion models can benefit homeland security. These models have the potential to greatly assist emergency management personnel in the:

- *preparedness stage* of predicting the outcome of C/B/N release scenarios;
- *response stage* of evaluating and containing the hazard zone; and
- *recovery and analysis stage* of assessing impacts on health and the environment.

Different dispersion modeling capabilities are required for each of these stages. The preparedness stage may include site-specific meteorological data coupled with probability-based dispersion model predictions and/or wind-tunnel simulations for typical scenarios. During the response stage, short execution time dispersion models are essential for providing emergency personnel with event-specific forecast data. During the recovery stage, all available data can be incorporated into a dispersion model designed to reconstruct the plume's space/time concentration distribution.

Dispersion models, particularly in the response and recovery stages, require meteorological observations to initialize the local wind field and contaminant data and, in turn, to initialize the source characteristics. Surface characteristics (e.g., topography, vegetation, built environment) for the area upwind and encompassing the C/B/N impact zone are also an important model input. All of these model inputs may vary from simplistic to highly complex, depending on the sophistication of the dispersion model. An additional challenge is that atmospheric dispersion models must be capable of assimilating measurements that come from an assortment of data collection networks, with information of uneven quality and quantity, collected over irregular time periods. For a model to be useful in the response stage of C/B/N events, input data must be available in real time and the model must have a short execution time.

The predicted concentration field from a dispersion model is combined with source toxicity, persistence, human and environmental sensitivity factors, and geographical information data to create maps of the event impacts. These maps are critical in the efficient allocation of emergency resources in the preparedness, response, and recovery and analysis stages of C/B/N events. To meet the needs of emergency response personnel, a dispersion model should map the hazard zone and provide an estimate of the concentration or dosage PDF at locations throughout the plume's domain.

The accuracy of a dispersion model's output (a statistical description of concentration in space and time) will depend on the quality of model inputs, the model's analytical methodology, and the inherent random nature of turbulent processes in the atmosphere. As discussed earlier in this chapter, the "true" concentration field of a specific C/B/N event cannot be predicted. However, a probabilistic description of the concentration field can be estimated via dispersion modeling, even with an incomplete wind field input.

Hazard Source Characterization

The C/B/N source characteristics (location, release rate, timing, buoyancy, momentum, toxicity, persistence, etc.) are critical in defining the ultimate event impact. This potential source variability requires that dispersion models include scales of motion ranging from meters to thousands of kilometers and account for chemical reactions and particle deposition physics.

Figure 4.6 depicts a typical decision process for C/B/N source characterization. If the source is a known hazard, then remote detection of source character can be automatic with proper pre-event planning. If an unknown source is imported into a likely target area, then remote real-time instrumentation may yield sufficient data to initialize the dispersion model. In the case of an unknown source released in an area with little real-time instrumentation, trained first-response personnel with portable sensors must define the source character (hospitals and local or regional poison centers may be able to provide additional information through symptom identification) for subsequent dispersion modeling. Defining the source quickly and accurately is extremely important for the successful application of a dispersion model in the response stage of C/B/N events. To respond effectively to a weapon release, remote observational instrumentation coupled with source prediction algorithms must have been implemented previously.

Decisions about the source toxicity and persistence will determine the type of transport and dispersion model most suitable for C/B/N events. Significantly different transport and dispersion methodologies and observational data requirements will be employed based upon the anticipated extent of the hazard zone.

Wind Field Characterization

Depending on the sophistication of the dispersion model, the availability of real-time data, and the horizontal scale of the area over which the dispersion must be calculated, the *wind field* may be defined by a variety of methods. These include using:

• a mean wind vector with an estimate of atmospheric stability in the vicinity of the release;

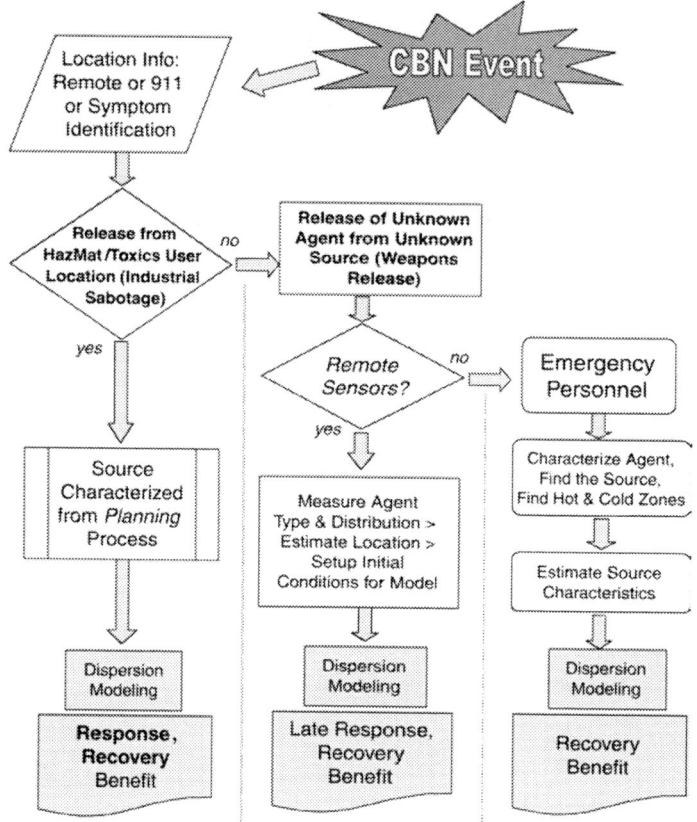

FIGURE 4.6 Hazard source characterization time line.

- a spatial array of winds provided by an analysis system that employs mass-continuity constraints; and
- a spatial array of winds and other meteorological parameters provided by a low-resolution Computational Fluid Dynamics (CFD) model or a mesoscale model that ingests observations.

As higher speed computer processing becomes available, high-resolution model simulations of the wind and turbulence will be available on the fine scales of urban canyons. However, presently full-physics simulations are limited to the large scales of metropolitan areas. Finer scales must be estimated using systems with more limited physics.

These observational wind data are processed in some dispersion models with local surface characteristics (topography, vegetation, structures) to form an estimate of the spatial and temporal wind field over the domain of C/B/N events. Lagrangian models use this wind field to transport and disperse either particles or Gaussian puffs to form concentration predictions. Figure 4.7 is an example using this type of system. CFD[4] models and wind-tunnel models require the wind field and the surface characteristics as initial and boundary conditions prior to simulation. The CFD model may calculate hazard concentrations with each time step (coupled solution) or it may solve for only the wind field, which subsequently is used in a Lagrangian tracking model (uncoupled solution). CFD models for most turbulent flow problems fall into two general categories: Reynolds-averaged Navier Stokes (RANS) and LES. RANS uses the ensemble-mean equations of motion, and LES uses the spatially averaged equations with spatial resolution adequate to resolve the largest-scale turbulent eddies in the flow field.

The typical spatial ranges of several dispersion modeling methods are depicted in Figure 4.8. The geographic extent of the wind data used for dispersion modeling should be several times greater than the anticipated maximum extent of the hazard (i.e., if the level of concern will stay within an urban area, then wind data in the surrounding area

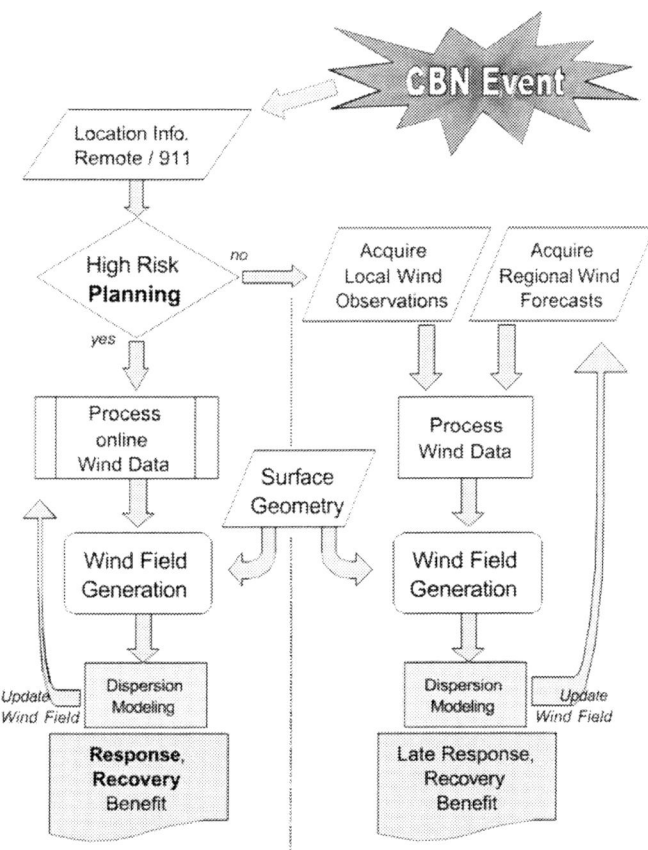

FIGURE 4.7 Wind field characterization time line.

---

[4] CFD is a numerically based solution technique that solves the governing conservation equations for fluid transport physics. The solution provides flow values (velocity, pressure, temperature, concentration, etc.) at a large number of grid points within a predefined domain.

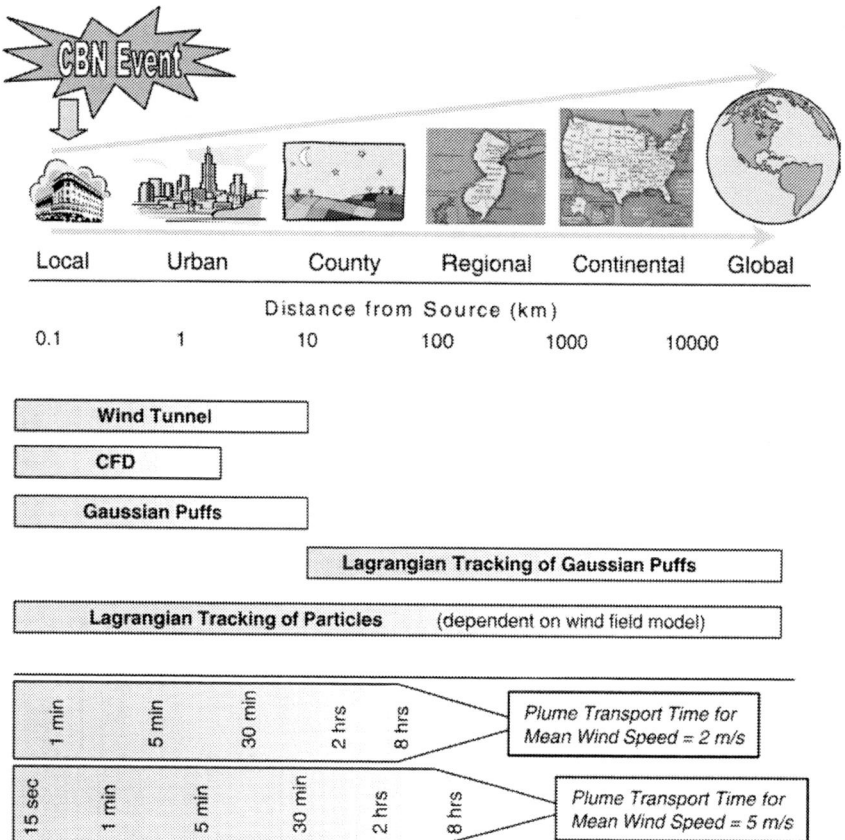

FIGURE 4.8 Spatial range of various dispersion model types.

upwind of the urban area are required). Figure 4.8 also shows the relationship between plume arrival time and the distance from the source for mean wind velocities of 2 and 5 ms^{-1}. In the case of urban C/B/N events, the plume likely will have passed through the entire urban zone in the early stages of event response (i.e., the first hour after the release).

REVIEW OF SELECTED C/B/N DISPERSION MODELING SYSTEMS

During the workshop proceedings, several presentations discussed the current status of selected dispersion modeling systems that were applicable to C/B/N events. A summary of these modeling systems is presented in Table 4.1, and all acronyms are defined in a list at the end of the report.

CAMEO, HPAC, and NARAC are operational quick-response systems that are in use today, each serving a separate user base. All three modeling systems have modules for source identification (e.g., chemical, biological, and/or nuclear databases), meteorological data input, dispersion modeling, and consequence analysis with graphic output. The dispersion models in each of these modeling systems are unique.

The Computer-Aided Management of Emergency Operations (CAMEO) system uses the Areal Locations of Hazardous Atmospheres (ALOHA) dispersion model, which is a modified Gaussian-plume formulation that predicts ensemble-averaged concentrations (also time averaged to several minutes) out to a distance of 10 km. Wind data from only one meteorological station are used in this model. For wind speeds of less than 1 ms^{-1}, it draws a wind-direction-independent envelope around the source. If the source gas is heavier than air, it uses a modification of the Dense Gas Dispersion Model (DEGADIS), a freeware PC program that executes rapidly but has no provisions for topography or individual building geometry.

The Hazard Prediction and Assessment Capability (HPAC) system has chemical, biological, and nuclear databases for source identification purposes, and it accesses weather data from in-house, NWS, and military providers. HPAC uses the SCIPUFF dispersion model, which uses a collection of Gaussian puffs to predict both the ensemble average concentration and the concentration variance out to regional scales. For dispersion distances less than approximately 10 km, its output characteristics are similar to the ALOHA model described above. The local topography is incorporated into the model via generation of an interpolated wind field with data from any number of surface and upper-air measurements, but the surface roughness is required to be constant over the entire domain. If the source gas is heavier than air, it uses a modification of the DEGADIS dense gas dispersion model. It is a registered user freeware PC program that has moderate execution times, but it has no provisions for individual building geometry.

Similar to HPAC, the National Atmospheric Release Advisory Center (NARAC) system has all three databases for source identification, and it also receives weather data from in-house, NWS, and military providers. NARAC uses a suite of dispersion models to custom tailor event predictions to a subscribing client's needs. The system runs 24 hours a day, 7 days a week so that near-real-time mockups of release events are possible via a network link. The dispersion models in NARAC range from a simple Gaussian puff model (INPUFF), to Lagrangian particle methods (LODI), to CFD approaches (FEM). Specifically for nuclear and chemical applications, NARAC has the stand-alone (non-reachback) Gaussian-plume models HOTSPOT and EPICode. The only building-aware[5] models in the NARAC system are CFD-based and have slow computation times. Both HPAC and NARAC models have been tested in the URBAN 2000 field experiment (Allwine et al., 2002).

Los Alamos National Laboratory (LANL) is in the final stages of testing two dispersion models that are designed for predicting hazardous concentrations in the urban environment. One is a Lagrangian particle dispersion model (QWIC-PLUME) coupled with a diagnostic wind field model (QWIC-URB). The other is a CFD–LES model named HIGRAD. Both models are building aware and were compared to URBAN 2000 field data.

---

[5] Building aware means that the model does consider individual building geometry.

TABLE 4.1 Models referenced in this report along with their corresponding description and features.

Sponsoring Agency	Model Acronyms	Model Description[a]	Operational Readiness	User Base	Execution Speed[b]	Domain Extent[c]	Terrain Aware[d]	Building Aware[e]	Concentration Methodology[f] (coordinate system/type of averaging)	Wind Field Methodology[g]	Concentration Output Type[h]			
											Ensemble Outputs			Space Avg
											Mean	Var	PDF	
NOAA/EPA	CAMEO/ ALOHA	Gaussian and heavy gas	Available for PC-based real time use	First responder and planner	Real time	Local to urban	No	No	Eulerian–ensemble	Single station input	Yes	No	No	—
DTRA/DOD	HPAC/ SCIPUFF	Gaussian puff heavy gas	Secure network	Military and civilian	Medium	Local to regional	Yes	No	Eulerian–ensemble	Diagnostic	Yes	Yes	No	—
LLNL/DOE	NARAC/ ADAPT-LODI	Lagrangian particle	Both open and secure systems	DOE, DOD, federal agency EOC, local agencies	Fast	Urban to regional to global	Via wind field	No	Lagrangian	Dynamical equation and diagnostic	Yes	No	No	Yes
	NARAC/ FEM3MP	CFD–RANS	Internal use	DOE	Medium to slow	Local to urban to regional	Via B.C.	Yes	Eulerian–ensemble	Dynamical equation	Yes	No	No	Yes
	NARAC/ FEM3MP	CFD–LES	Internal use	DOE	Slow to very slow	Local to urban to regional	Via B.C.	Yes	Eulerian–spatial	Dynamical equation	Multi-runs	Multi-runs	Multi-runs	Yes
LANL/DOE	QWIC	Lagrangian particle	Final testing not deployed	DOE	Fast	Local to urban	Via wind field	Yes	Lagrangian	Diagnostic	Yes	No	No	Yes
	HIGRAD	CFD–LES	Final testing not deployed	DOE	Very slow	Local to regional	Via B.C.	Yes	Eulerian–spatial	Domain inlet profile	Multi-runs	Multi-runs	Multi-runs	Yes
EPA	CALPUFF	Gaussian puff	Available for use	Freeware	Medium	Local to regional	Via wind field	No	Eulerian–ensemble	Diagnostic	Yes	No	No	—

	Model Description[a]			Execution Speed[b]	Domain Extent[c]	Terrain Aware[d]	Building Aware[e]	Concentration Methodology[f]	Wind Field Methodology[g]				Concentration Output Type[h]
	Fluent®	CFD–RANS and LES	Internal use	Commercial	Slow	Local to urban	Via B.C.	Yes	Eulerian–ensemble	Domain inlet profile	Yes	No	No	Yes
	Wind Tunnel	Reduced scale physical model	Internal use	Case studies	Very slow	Local to urban	Yes	Yes	Eulerian–spatial	Developed boundary layer	Multi-runs	Multi-runs	Multi-runs	Yes
SAIC	OMEGA	Lagrangian particle	Available for use	Client subscription	Medium	Local to global	Via B.C.	No	Lagrangian	Dynamical equation	Yes	Yes	No	Yes
UK DSTL	UDM	Gaussian puff	Secure network	UK–Defense	Fast	Local to regional	Via wind field	Yes	Eulerian–ensemble	Diagnostic	Yes	No	No	-

[a] Model Description: Lagrangian particle refers to a Lagrangian model of the ensemble-average turbulent diffusion; RANS refers to turbulence models based on equations for ensemble-averaged turbulent fluxes; LES refers to the large eddy simulation CFD method; Gaussian puff refers to a sequence of instantaneous releases that use an empirically based growth model.
[b] Execution Speed: Real time=less than 5–10 seconds; fast=less than 5–10 minutes; medium=less than 2–4 hours; slow=less than 24 hours; very slow=greater than 24 hours.
[c] Domain Extent: See Figure 4.8 for definitions.
[d] Terrain Aware: "via wind field" means a separate wind field model has created flow vectors that satisfy the continuity equation over the domain's terrain; "via B.C." means that the CFD model considers terrain effects through its boundary conditions.
[e] Building Aware: Does the model consider individual building geometry?
[f] Concentration Methodology: First entry is the type of coordinate system used; second entry is the type of averaging used in Eulerian models.
[g] Wind Field Methodology: "Diagnostic" is a wind field that is interpolated from observed data; "dynamical equation" is a wind field that is calculated from the dynamical equations using observed data as initial conditions.
[h] Concentration Output Type: "Var" stands for variance; "PDF" stands for probability density function; "multi-runs" means that an ensemble average of a statistical quantity can be obtained via multiple runs with different initial conditions; "space averaging" is dependent on the model's grid spacing or for a wind-tunnel model's measurement volume.

The LANL presentation (Appendix I) provided a concentration comparison between the URBAN 2000 field measurement program and the United Kingdom Defense Science and Technology Laboratory urban dispersion model (UDM). This Gaussian-puff model[6] was specifically designed for dispersion in an urban environment and has been evaluated in many field and wind-tunnel experiments.

The Environmental Protection Agency (EPA) presentation (Appendix F) demonstrated the use of several dispersion models to study the plume from the World Trade Center towers fire. EPA is developing a Gaussian puff dispersion model (CALPUFF), coupled with the diagnostic wind field model (CALMET); a CFD model (FLUENT); and a wind-tunnel simulation approach. The CALPUFF model produces one-hour averaged concentrations, thus it should be used only for steady sources. The CFD and wind-tunnel models are building aware, but the CALPUFF model is not.

The Science Applications International Corporation presented the Operational Multiscale Environmental model with Grid Adaptivity (OMEGA), which is capable of using either an Eulerian or a Lagrangian particle approach. In OMEGA, the dispersion is fully coupled to a high-resolution NWP system with extra emphasis on surface and boundary layer processes. OMEGA has been used to support short-range dispersion in complex terrain (e.g., White Sands Missile Range) and long-range dispersion at continental scale (e.g., the European Tracer Experiment, ETEX). The OMEGA presentation included comparisons of predicted and observed concentration for ETEX.

DISCUSSION OF C/B/N MODELING SYSTEMS

Dispersion model systems applied to C/B/N event scenarios can be divided into those that are useful for pre-event planning and training, those that are immediately available for response tactics, and those that will be used for post-event evaluation and recovery[7]. Dispersion models used for C/B/N event planning and response should provide emergency personnel with a common impact mapping format. In particular, given a dosage level of concern (LOC) for a toxin and a prediction confidence level, the dispersion model should provide a spatial contour defining the three-dimensional hazard zone. For example, if the confidence level was set at 99 percent, the dosage LOC would occur outside of the hazard zone contour only one time in 100 independent release events. To define the hazard zone in this format, some estimate of the spatial distribution of the concentration PDF is required. Several of the models discussed at the workshop do not currently provide sufficient statistical information to estimate PDFs. Models that cannot provide hazard zone confidence levels are of limited usefulness in emergency management; they are better suited to the chronic release situation of air pollution modeling.

CFD–LES models and laboratory simulations (i.e., reduced-scale models in a wind tunnel) of urban dispersion are the only tools currently available that can create the PDF

---

[6] This type of model divides emissions into a series of overlapping volumes (puffs), so that it is not necessary to assume horizontally homogeneous emissions or to require steady-state conditions.
[7] Note that post-event concentration data, if dense enough, could recreate the concentration field of an event.

of concentrations or dosages downwind of a transient C/B/N event. In both the wind-tunnel and the LES approaches, an ensemble of independent simulations covering a range of wind direction and wind speed may need to be formed in order to include scales of motion greater than that present in the domain of the model. This decoupling of turbulent motion scales will introduce errors in the model's predictive capabilities. No model in existence today can precisely deal with the full range of motions present in the urban dispersion problem; hence, there is a need for new computational tools to address this issue. Laboratory simulations are important tools for creating site-specific databases of C/B/N event scenarios and for the development and evaluation of both fast-response urban dispersion models and CFD-based dispersion models. Laboratory simulations provide better resolution of turbulent motions than current CFD models in the urban setting. CFD models have the potential to predict dispersive events in flow regimes that laboratory simulations find difficult, such as low wind speed and thermally dominated flows. CFD–LES modeling approaches also could potentially be used to study where in an urban area a plume of hazardous material likely is to be most heavily deposited. CFD models interface more interactively with meteorological data systems, even though execution times on the order of several days are common. With today's technology, a wind-tunnel urban simulation with enough data to define the probability density function throughout the plume's domain would take about a week to perform.

The time required to build site-specific urban boundary conditions for both the CFD and the laboratory simulations would be substantially reduced if each urban area of concern had three-dimensional databases of buildings and topography that were compatible with the dispersion modelers' needs. The National Imagery and Mapping Agency of the U.S. Geological Survey already has work underway in this area as part of its National Mapping Program (see http://mapping.usgs.gov/). These databases should be flexible in the amount of detail they provide so as to not overwhelm the computational model. Such databases would be useful for many other purposes, for example, air pollution modeling and urban planning for extreme winds.

To simplify dispersion models and reduce their error, it is desirable to define a minimum spatial and temporal scale at which averaged concentrations (over specific spatial and/or temporal scales) will be sufficient to determine C/B/N event impacts on health and the environment. This minimum scale will depend on the toxin released; thus, prior to designing a dispersion model, the potential range of toxin dosage levels of concern should be explored.

Because of the importance of proper preparation before sending emergency personnel into harm's way, it is prudent to conservatively predict the extent of hazard zones. This might be accomplished by using statistical data obtained through an ensemble of model runs and by setting zone thresholds conservatively, based on the statistics of the ensemble. Introducing real-time observations into the process through data assimilation or the comparison of model with observational data would increase confidence levels.

> **BOX 4.1**
> **Need for Improved Knowledge and Modeling of Urban Meteorology**
>
> The winds that transport C/B/N material in a city result from a superposition of motions, ranging from the continental scale to the scales of individual urban structures. These motions are defined by synoptic-scale weather systems, mesoscale wind systems that may be generated by local terrain or coastlines, and the aggregate weather effects of entire urban complexes (which include strong mechanical and thermal forcings that operate over limited areas). In order to provide accurate meteorological input to dispersion models, all of these scales need to be represented, either through observations or through simulation models. A hierarchy of modeling and observational capabilities may be needed for immediate estimation of the short-distance transport of a hazardous plume. For instance, a single undisturbed wind observation upwind of the urban complex may suffice to provide the general direction of plume movement. A few undisturbed upwind observations of the three-dimensional winds and stability may suffice to provide input to a model that diagnoses dynamic effects on the mean airflow of buildings and local topography. For longer-range transport, it is necessary to represent complex spatial and temporal variations in meteorological conditions. For this purpose, general meteorological models are needed. The model should be able to provide a detailed, physically consistent analysis of the current urban-scale meteorology, and it must also be able to predict the future state of the meteorology, ideally for as long as the plume remains hazardous.

It is clear that the existing suite of dispersion models currently in operational use by various government agencies has room for improvement. Both fast-execution response models and slower (but more accurate) preparedness and recovery models need further development and evaluation. Once a viable set of dispersion models capable of C/B/N event predictions is established, an independent quantitative review of these models should be initiated, and the results should be used to improve model performance. It ultimately may be determined that an ensemble of outputs from different models would yield better dispersion estimates than those of any one model alone.

Many of the intercomparison studies carried out to date have been qualitative in nature and lacking in carefully controlled ground rules. Simplistic attempts to compare models against one another may serve to validate those models that have been preferentially designed for the conditions prescribed by the given competition, but this does not necessarily indicate superior modeling approaches for other, more arbitrary conditions. What are needed are carefully designed intercomparison studies that allow quantitative evaluation of models under the same controlled conditions. Procedures will need to be formulated so that experiments and models with significantly different output formats (e.g., field experiments producing a single realization and model outputs of ensemble-average statistics) are properly evaluated. Such rigorous intercomparisons have never been done adequately to date and will require careful experimental design. Proper evaluation would be aided by full documentation of each model's range of applicability, typical setup and execution times, forms of output (e.g., ensemble or spatial averaging), analytical methods used for dealing with plume advection and growth and with different scales of motion, and other relevant factors.

KEY FINDINGS AND RECOMMENDATIONS

For purposes of threat assessment, preparation, and training, existing dispersion models meet some needs of the emergency response community. In the case of actual emergencies, the needs of emergency management may not be well satisfied by existing models. In particular, single-event uncertainties in atmospheric dispersion models are not well bounded, and current models are not well designed for complex natural topographies or built urban environments.

Most available atmospheric dispersion models predict only the ensemble-average concentration (that is, the average over a large number of realizations of a given dispersion situation). New approaches are needed for modeling a single hazardous release.

Dispersion models used for emergency planning and response should provide confidence estimates that prescribed concentrations will not be exceeded outside of predicted hazard zones. This requires that models provide some measure of the possible variability in a given situation.

Different dispersion modeling methodologies are required in the preparedness, response, and recovery stages of C/B/N events. For the preparedness stage, an accurate model capable of providing confidence-level estimates is desired, but model execution time is not important. For the response stage, accuracy can be compromised to obtain timely predictions, but the dispersion model must still provide confidence-level estimates. For the recovery stage, model execution time is not important, but accurate model reconstruction of the plume concentration distribution over time is desired. In order to use a dispersion model's predictions effectively during the early response phase, the wind field and other conditions at the site of the release must be available in near real time and a short model execution time is essential. The most appropriate dispersion model for any given scenario may depend on the quantity, toxicity, and persistence of the hazardous agent; thus, it is critical that source identification be as rapid as possible.

The committee's review of selected existing dispersion modeling systems determined that no one system had all the features that the committee deemed critical: confidence estimates for the predicted dosages, accommodation of urban and complex topography, short execution time urban models for the response phase, and accurate though slower models for the preparedness and recovery phases. Better integration between existing and future modeling systems could supply all of these critical features.

The "unpairing" of concentration predictions and observations in time and space (commonly done with continuous sources in air quality applications) is inappropriate when evaluating dispersion model performance in episodic releases. Evaluation techniques based on more advanced probabilistic methods need to be developed. Toward that end, existing dispersion models should identify the type of averaging (ensemble, time and space) inherent in their modeling methodology, both in the wind field formulation and in the treatment of dispersion. The reliability of existing and future dispersion modeling systems should be evaluated against field and laboratory measurements for potential C/B/N event scenarios. If predicted confidence limits are found to be unacceptable, then empirical corrections should be applied to model outputs so as not to place emergency personnel in harm's way. Increasing the density of the wind measurements in a plume's domain will potentially reduce uncertainty, thus reducing the predicted extent of the hazard without compromising confidence.

Meteorological observations are a critical element of dispersion modeling. Observational technologies have been evolving rapidly in recent decades, and the committee has identified many existing measurement technologies that have not been fully exploited through data assimilation. Model operators and developers would benefit from broader interaction with the meteorological community, to take advantage of leading-edge research in data assimilation, quantitative precipitation forecasting, short-range numerical weather prediction, and high-resolution forecasting initialized with radar data. Likewise, observational research programs studying issues such as weather prediction, properties of boundary layer turbulence, and air pollution transport should be viewed as targets of opportunity for testing and evaluating dispersion models.

A nationally coordinated effort should be established to foster support and systematic evaluation of existing models and research and development of new modeling approaches, undertaken in collaboration with the broader meteorological community. The Office of the Federal Coordinator for Meteorology, which recently organized a review of U.S. dispersion modeling capabilities, could provide valuable input as to which agency(ies) is best suited to oversee this coordinated effort. Among the issues that should be addressed through this coordinated program are the following:

- New dispersion modeling constructs need to be further explored and possibly adapted for operational use in urban settings. This includes advanced, short execution time models; slower but more accurate computational fluid dynamics and large-eddy simulation models; and models with adaptive grids.
- Techniques must be developed for constructing ensembles of model solutions on the urban scale so that probabilistic rather than deterministic information can be provided to emergency managers. It will be necessary to quantify the level of confidence as a function of the number of ensemble members, which in turn, will have implications for the computational power required.
- It is necessary to learn how to more effectively assimilate into models an appropriate range of meteorological data (e.g., wind, temperature, and moisture data) from observing systems as well as real-time data from C/B/N sensors, especially as the quality and availability of these data increase. It also is important to effecttively couple dispersion models with appropriate source characterization models.
- Urban field programs and wind-tunnel simulations should be conducted to allow for the testing, evaluation, and development of existing and new modeling systems (both meteorological and dispersion models). Developing an appropriate experimental design for such studies is a critical task that itself will require careful evaluation.

- The bulk effects of urban surfaces on the surface energy, moisture, and momentum are not well accounted for in most meteorological models. Existing development work in this area should be enhanced, and the improved modeling techniques adopted more widely.
- Urban building and topography three-dimensional databases need to be developed and maintained for use in numerical and wind-tunnel dispersion simulations.

In at least one large urban area, a fully operational dispersion tracking and forecasting system should be established—that is, a comprehensive system for collecting relevant meteorological and C/B/N sensor data, assimilating this information into a dispersion model, and maintaining the expertise and organizational capacity to provide immediate model forecasts on a full-time basis. If possible, a few such systems should be established and evaluated for different types of urban areas (e.g., coastal versus continental cities, low-latitude versus high-latitude cities). Such systems can be used as test beds for gaining understanding of model capabilities and limitations, and their use should not be limited to emergency situations. These observational and modeling tools could have multiple applications, which would help justify costs and ensure that the systems are frequently used, maintained, evaluated, and quality controlled.

There is a wealth of knowledge about meteorological and dispersion models residing in universities, NWS Weather Forecast Offices, and private sector facilities throughout the nation. These sources of expertise, together with the existing programs in several national laboratories and military facilities, should be integral components of the coordinated national effort recommended above, to assist with developing local and regional models that are optimized for the topography and seasonal weather patterns in vulnerable areas. At the most basic level, this integration can be implemented via collaborative research and development efforts.

BOX 4.2
Management and Coordination Needs

There is a wide array of federal agencies that operate dispersion modeling systems, including the Department of Commerce–NOAA, Department of Defense, Department of Energy, EPA, Federal Emergency Management Agency, and Nuclear Regulatory Commission, along with numerous academic and private sector research groups that contribute to these federal efforts. In addition, it must be recognized that the new Department of Homeland Security, established in January 2003, may eventually augment or subsume some of the activities and responsibilities currently residing in these other federal agencies. At the present time, however, it is not known to the committee what specific organizational plans are being considered.

Given the ambiguity of this situation and the limited time and resources available to examine these management-related issues, the committee felt that it was not appropriate to make specific suggestions about agency leadership responsibilities for the various activities recommended in this report. The committee emphasizes, however, that a carefully crafted management strategy, with clear lines of responsibility and authority, is essential for ensuring further progress in the development and ongoing operation of dispersion modeling systems. There is a clear need for more central coordination among the various federal agencies currently involved and among the relevant players at local, regional, and national levels.

Each of the agencies mentioned above has developed its own "customer base" and areas of strength and specialization; thus, it seems likely that some form of distributed responsibility will continue to be the most effective organizational strategy. However, a strong center of coordination is needed to ensure that the necessary research and development work is carried out and that emergency responders have unambiguous guidance as to where to turn for help.

References

Allwine, K.J., J.H. Shinn, G.E. Streit, K.L. Clawson, and M. Brown. 2002. Overview of URBAN 2000: A multiscale field study of dispersion through an urban environment. *Bull. Amer. Meteor. Soc.* 83:521-551.

Banta, R.M., L.S. Darby, P. Kaufman, K.H. Levinson, and C.J. Zhu. 1999. Wind flow patterns in the Grand Canyon as revealed by Doppler lidar. *J. Appl. Meteor.* 38:1069-1083.

Bosart, L.F. 1983. Analysis of a California Catalina eddy event. *Mon. Wea. Rev.* 111:1619-1633.

Brock, F.V., K. Crawford, R.L. Elliott, G.W. Cupserus, S.J. Sadler, H.L. Johnson, and M.D. Eilts. 1995. The Oklahoma Mesonet: A technical overview. *J. Atmos. Ocean. Tech.* 12:5-19.

Chang, J.C., P. Franzese, K. Chayantrakom, and S. Hanna. 2003. Evaluations of CALPUFF, HPAC, and VLSTRACK with two mesoscale field datasets. *J. Appl. Meteor.* 42:453-466.

Darby, L.S., K.J. Allwine, and R.M. Banta. 2002. Relationship between tracer behavior in downtown Salt Lake City and basin-scale wind flow. Pp. 12-15 in *Proc. 10th Conf on Mountain Meteorology and MAP meeting 2002*, Park City, Utah, June 17-21.

Martner, B.E., D.B. Wuertz, B.B. Stankov, R.G. Strauch, E.R. Westwater, K.S. Gate, W.L. Ecklund, C.L. Martin, and W.F. Dabberdt. 1993. An evaluation of wind profiler, RASS, and microwave radiometer performance. *Bull. Amer. Meteor. Soc.* 74:599-613.

Mayor, S.D., R.J. Alvarez II, C. Senff, R.M. Hardesty, C.L. Frush, R.M. Banta, and W.L. Eberhard. 1996. *Second International Airborne Remote Sensing Conference and Exhibition*, San Francisco, California, June 24-27.

Mecikalski, J.R., D.B. Johnson, and J.J. Murray. 2002. NASA Advanced Satellite Aviation Weather Products (ASAP) Study Report. NASA Technical Report, in press.

Mesonet. 2002. Abstracts from the Mesonet 2002 Conference, Oklahoma City, June 23-26. (Sponsored by the Oklahoma Climatological Survey, the Oklahoma Mesonet, and a grant from Innovations in American Government, sponsored by the Ford Foundation and awarded through the Institute for Government Innovation at Harvard's John F. Kennedy School of Government.)

Miller, J. (2003, January 22). Threats and Responses: Biological Defenses; U.S. Deploying Monitor System for Germ Peril. *The New York Times*, pp. A1.

Moninger, W.R., R.D. Mamrosh, and P.M. Pauley. 2003. Automated meteorological reports from commercial aircraft. *Bull. Amer. Meteor. Soc.* 84:203-216.

Morris, D.A., K.C. Crawford, K.A. Kloesel, and J.M. Wolfinbarger. 2001. OK-FIRST: A meteorological information system for public safety. *Bull. Amer. Meteor. Soc.* 82:1911-1923.

NRC (National Research Council). 1995. *Toward a New National Weather Service—Assessment of NEXRAD Coverage and Associated Weather Services.* National Academy Press, Washington, D.C.

NRC. 2002a. *National Security and Homeland Defense: Challenges for the Chemical Sciences in the 21st Century.* National Academy Press, Washington, D.C.

NRC. 2002b. *Weather Radar Technology Beyond NEXRAD.* National Academy Press, Washington, D.C.

NRC. 2002c. *Making the Nation Safer: The Role of Science and Technology in Countering Terrorism.* National Academy Press, Washington, D.C.

Rothermal, J., D.R. Cutten, R. M. Hardesty, R.T. Menzies, J.N. Howell, S.C. Johnson, D.M. Tratt, L.D. Olivier, and R.M. Banta. 1998. The multi-center airborne coherent atmospheric wind sensor. *Bull. Amer. Meteor. Soc.* 79:581-599.

Shepherd, J.M., and A.V. Mehta. 2002. *Summary of First GPM Partners Planning Workshop.* NASA Conference Publication-2002-210012-GPM Report 1. Available at NASA-Goddard Space Flight Center, 1-37.

Shepherd, J.M., and E.A. Smith. 2002. *Bridging from TRMM to GPM to 3-Hourly Precipitation Estimates.* NASA Tech. Memo.-2002-211602-GP—Report 7. Available at NASA-Goddard Space Flight Center, 1-7.

Weil, J.C., R.I. Sykes, and A. Venkatram. 1992. Evaluating air-quality models: Review and outlook. *J. Appl. Meteorol.* 31:1121-1145.

Weil, J.C., L.A. Corio, and R.P. Brower. 1997. A PDF dispersion model for buoyant plumes in the convective boundary layer. *J. Appl. Meteorol.* 36:982-1003.

Wilczak, J.M., and J.W. Glendenning. 1988. Observations and mixed-layer modeling of a terrain-induced mesoscale gyre: The Denver Cyclone. *Mon. Wea. Rev.* 116:1599-1622.

Acronyms and Abbreviations

ADAPT	Atmospheric Data and Parameterization Tool – NARAC
AERMIC	AMS–EPA Regulatory Model Improvement Committee
AERMOD	AERMIC Dispersion Model
AFTOX	U.S. Air Force Toxic Dispersion Model
ALOHA	Areal Locations of Hazardous Atmospheres – EPA/NOAA
ANATEX	North America Tracer Experiment
ARPS	Advanced Regional Prediction System model
BASC	Board on Atmospheric Sciences and Climate
BT	Burk–Thompson parameterization
CALMET	California puff Meteorology module – EPA
CALPUFF	California Puff dispersion module – EPA
CAMEO	Computer-Aided Management of Emergency Operations – EPA/NOAA
CAPTEX	Cross Appalachian Tracer Experiment
CATS-JACE	Consequence Assessment Tool Set–Joint Assessment of Catastrophic Events – DTRA
C/B/N	chemical/biological/nuclear
CFD	Computational Fluid Dynamics
CMAQ	Community Multiscale Air Quality Model – EPA
COAMPS	Coupled Ocean-Atmospheric Mesoscale Prediction System – Naval Research Laboratory
CRSTER	a standard Gaussian-plume model – EPA
DEGADIS	Dense Gas Dispersion Model – EPA
DOD	Department of Defense
DOE	Department of Energy
DP26	Dipole Pride 26 field experiment
DTRA	Defense Threat Reduction Agency
ECMWF	European Center for Medium-Range Weather Forecasting
EPA	Environmental Protection Agency
EPIcode	Emergency Prediction Information code Gaussian-plume model for chemical releases – NARAC/LLNL
ETEX	European Tracer Experiment
FEM	Finite Element Model 3 – NARAC/LLNL

FLUENT	Commercial CFD model
HIGRAD	High resolution model for strong Gradient applications – LANL CFD model
HOTSPOT	Gaussian-plume model for radiological releases – NARAC/LLNL
HPAC	Hazard Prediction and Assessment Capability – DTRA
HYSPLIT	Hybrid Single-Particle Lagrangian Integrated Trajectory Model
INPUFF	Integrated Gaussian-Puff dispersion model – NARAC/LLNL
IOP	Intensive Operations Period
IR	infrared
JEM	Joint Effects Model
K	eddy diffusivity coefficient
LANL	Los Alamos National Laboratory
LES	large-eddy simulation technique used in CFD model
LLNL	Lawrence Livermore National Laboratory
LOC	(dosage) level of concern
LODI	Lagrangian Operational Dispersion Integrator particle dispersion model – NARAC
LSM	Land-Surface Model
MATHEW/ADPIC	Mass-Adjusted Three-Dimensional Wind Field/Atmospheric Dispersion by Particle-in-Cell
MDCRS	Meteorological Data Collection and Reporting System
MDT	mountain daylight time
MIDAS-AT	Meteorological Information and Dispersion Assessment System–Anti-Terrorism
MM5	Pennsylvania State University–NCAR mesoscale model
MRF	Medium-Range Forecast
MUST	Mock Urban Settings Test field experiment
NARAC	National Atmospheric Release Advisory Center
NCAR	National Center for Atmospheric Research
NCEP	National Centers for Environmental Prediction
NEXRAD	Next Generation Radar
NOAA	National Oceanic and Atmospheric Administration
NOGAPS	Navy Operational Global Atmospheric Prediction System
NRC	National Research Council
NWP	Numerical Weather Prediction
NWS	National Weather Service
OBDG	Ocean Breeze Dry Gulch model
OLAD	Overland Atmospheric Dispersion field experiment
OMEGA	Operational Multiscale Environmental model with Grid Adaptivity – SAIC
PATRIC	Particle Trajectory-in-Cell model
PDF	Probability Density Function of a random variable
Q-Q	Quantile-Quantile
QWIC-PLUME	Fast Urban Dispersion Model, Dispersion Module – LANL
QWIC-URB	Fast Urban Dispersion Model, Meteorological Module – LANL
RAMS	Regional Atmospheric Modeling System
RANS	Reynolds Averaged Navier-Stokes equation used in CFD model
RASS	Radio Acoustic Sounding System

RODOS	Real-time Online Decision Support System
SAFER	Safety Assessment for Explosives Risk
SAIC	Science Applications International Corporation
SCIPUFF	Second-order Closure Integrated Puff model – EPA
TKE	turbulent kinetic energy
UAV	unmanned aerial vehicle
UHF	ultra-high frequency
URBAN 2000	Field campaign to study the urban environment and its effect on atmospheric dispersion
VLSTRACK	Chemical/biological agent Vapor, Liquid and Solid Tracking model – DOD
VTMX	Vertical Transport and Mixing Experiment
WSR-88D	Weather Surveillance Radar-1998 Doppler

Committee Biographies

ROBERT J. SERAFIN (chair) is Director Emeritus of the National Center for Atmospheric Research. His technical interests are related to radar, remote sensing, and in situ sensing of the atmosphere. He has expertise in the areas of signal processing theory, Doppler radar, lidar, and passive remote sensing techniques, and in the use of such systems for applications including severe weather detection, weather forecasting, precipitation estimation, and hydrological studies. Dr. Serafin is a member of the National Academy of Engineering and a fellow of the American Meteorological Society and the Institute of Electrical and Electronics Engineers.

ERIC J. BARRON is Dean of the College of Earth and Mineral Sciences and Distinguished Professor of Geosciences at Pennsylvania State University. His professional experience encompasses scientist at the National Center for Atmospheric Research, associate professor of marine geology and geophysics at the University of Miami and director of the Earth System Science Center. His specialty is paleoclimatology and paleoceanography. His research emphasizes global change, specifically numerical models of the climate system and the study of climate change throughout Earth history. Dr. Barron is a fellow of the American Geophysical Union and the American Meteorological Society. He currently serves as chair of the NRC Board on Atmospheric Sciences and Climate.

HOWARD B. BLUESTEIN is Professor of Meteorology at the University of Oklahoma and a Noble Foundation Presidential Professor. He received his Ph.D. in meteorology from the Massachusetts Institute of Technology. His research interests are the observation and physical understanding of weather phenomena on convective, mesoscale, and synoptic scales. Dr. Bluestein is a fellow of the American Meteorological Society (AMS) and the Cooperative Institute for Mesoscale Meteorological Studies. He is past chair of the National Science Foundation's Observing Facilities Advisory Panel and a past member of the AMS Board of Meteorological and Oceanographic Education in Universities. He is also the author of a textbook on synoptic-dynamic meteorology. He is a member of the NRC Board on Atmospheric Sciences and Climate.

STEVEN F. CLIFFORD is a Senior Scientist at the Cooperative Institute for Research in Environmental Sciences and the former director of the NOAA Environmental Technology Laboratory. He received his Ph.D. in engineering science from Dartmouth College. His research goals include developing a global observing system using ground-based, airborne, and satellite remote sensing systems, and using these observations as input to global air-sea circulation models for improving forecasts of weather and climate change. He is a fellow of the Optical and Acoustical Societies of

America, a senior member of the Institute of Electrical and Electronics Engineers, and a member of the American Physical Society, the American Geophysical Union, and the American Meteorological Society. He is also a member of the National Academy of Engineering and the NRC Board on Atmospheric Sciences and Climate.

LEWIS M. DUNCAN is Dean of the Thayer School of Engineering at Dartmouth College. A space and plasma physics researcher for more than two decades, he specializes in high-power radiowave propagation and its applications in remote sensing and telecommunications. He has served as head of the Division of Earth and Space Sciences at Lawrence Livermore National Laboratory. Dr. Duncan has conducted public policy research on arms control, nuclear non-proliferation, counterterrorism and emerging threats, and he recently played a key role in bringing the new Institute for Security Technology Studies to Dartmouth.

MARGARET A. LEMONE is a senior scientist at the National Center for Atmospheric Research. Her scientific interests are in the structure and dynamics of the atmosphere's planetary boundary layer and its interaction with the underlying surface and clouds, and the interaction of mesoscale convection with the boundary layer and the surrounding atmosphere. Dr. LeMone is a member of the National Academy of Engineering and a fellow of the American Association for the Advancement of Science and the American Meteorological Society. She has served on the NRC's Panel on Improving the Effectiveness of U.S. Climate Modeling, and she currently is a member of the National Academy of Engineering's Program Development Committee and the NRC Board on Atmospheric Sciences and Climate.

DAVID E. NEFF is a Senior Research Scientist in Civil Engineering and Co-director of the Wind Engineering and Fluids Laboratory (WEFL) at Colorado State University (CSU). He received his Ph.D. in Civil Engineering from CSU, and his dissertation focused on physical modeling of heavy plume dispersion. His research projects include the physical and numerical modeling of wind turbine wake interactions; air quality prediction of pollutant recirculation in residential and industrial ventilation systems; air quality prediction of pollutant emissions from fossil fuel and nuclear power plant complexes; and hazard assessment from the accidental release of dense gases. Dr. Neff played a key role in the development and maintenance of CSU's WEFL experimental facility. He is a member of the American Association for Wind Engineering.

WILLIAM E. ODOM is a Senior Fellow and Director of National Security Studies at Hudson Institute's Washington, D.C. office. He is also an adjunct professor at Yale University. As Director of the National Security Agency from 1985 to 1988, he was responsible for the nation's signals intelligence and communications security. From 1981 to 1985, he served as Assistant Chief of Staff for Intelligence, the Army's senior intelligence officer. From 1977 to 1981, General Odom was Military Assistant to the President's Assistant for National Security Affairs, Zbigniew Brzezinski. As a member of the National Security Council staff, he worked on strategic planning, Soviet affairs, nuclear weapons policy, telecommunications policy, and Persian Gulf security issues. He is a graduate of the United States Military Academy and received his Ph.D. from Columbia University.

GENE J. PFEFFER works as a private consultant. He recently served as Colorado Springs Director of Orbital Sciences Corporation's Office of Defense Programs, where he dealt with a wide range of activities related to satellite systems and environmental sensors. Prior to this he was a consultant to the Federal Coordinator for Meteorological Services, and served for three decades in the

U.S. Air Force. His extensive military career includes serving as Vice Commander of the Air Force Weather Service and Director of Weather for the North American Aerospace Defense Command. He is a fellow of the American Meteorological Society, and he chaired their Board on Private Sector Meteorology. Mr. Pfeffer holds Masters degrees in Meteorology and in Systems Management.

KARL K. TUREKIAN is Silliman Professor of Geology and Geophysics at Yale University and Director of the Yale Institute for Biospheric Studies. His research areas include atmospheric geochemistry of cosmogenic, radon daughter and man-made radionuclides, surficial and groundwater geochemistry of radionuclides marine geochemistry and the study of climate change over geologic time. Dr. Turekian is a member of the National Academy of Sciences. He has served as Editor of several major scientific journals and was on the Editorial Board of the *Proceedings of the National Academy of Sciences*.

THOMAS J. WARNER is a Professor in the Program in Atmospheric and Oceanic Sciences at the University of Colorado, and a Scientist at the National Center for Atmospheric Research. His research interests include numerical modeling of mesoscale atmospheric phenomena and dynamics (with special emphasis on hydrologic processes), regional climate modeling, and data assimilation. A recent research emphasis has been the use of coupled atmospheric dynamic and dispersion models for estimation of the effects of domestic and international releases of hazardous material. He served on the NRC Panel on Model-Assimilated Data Sets for Atmospheric and Oceanic Research and was a speaker at a recent Congressional Forum on "National Security and the Atmosphere".

JOHN C. WYNGAARD is a Professor of Meteorology, Mechanical Engineering, and Geoenvironmental Engineering at Pennsylvania State University. He studies turbulence in the atmosphere through direct observations and supercomputer simulation. He is interested in new observational approaches, including ground-based remote sensing, as well as measurements from towers and aircraft, and he studies the dynamic performance of turbulence sensors. Using the large-eddy simulation technique, he is developing new representations of turbulence effects in meteorological and oceanographic models of local to global scales. Dr. Wyngaard is a member of the NRC Board on Atmospheric Sciences and Climate.

Appendixes

The material in these appendixes is aimed at providing the reader with more information about the workshop that served as the primary information-gathering focus for this report. Included are a copy of the workshop agenda and a list of participants (A). Also included are summaries of several workshop presentations, including three presentations that provide a general overview of the key discussion topics (B, C, and D), and five presentations that describe specific examples of how dispersion modeling systems can be applied in a "real world" context (E through I).

A

Workshop Agenda and Participant List

Tools for Tracking Chemical/Biological/Nuclear Releases in the Atmosphere: Implications for Homeland Security
July 22–23, 2002
Erik Jonsson Center, Woods Hole, Massachusetts

Monday, July 22, 2002

9:00 A.M.   Introductory Remarks (*Robert Serafin, committee chair*)

9:15 A.M.   Overview Talks[1]

- Atmospheric transport and dispersion modeling (*Steve Hanna, George Mason University*)
- Observations and data assimilation for atmospheric transport/dispersion studies (*Walter Dabberdt, Vaisala*)
- Information needs of emergency first responders (*Frances Edwards-Winslow, San Jose Emergency Preparedness Office*)

10:30 A.M.   Discuss examples of tools and programs currently employed (or in development) for modeling the dispersion of C/B/N agents[2], including:

- DOE's National Atmospheric Release Advisory Capability (NARAC) *(James Ellis, Gayle Sugiyama, LLNL)*
- Defense Threat Reduction Agency's Hazard Prediction and Assessment Capability *(Martin Bagley, Brian Beitler; DTRA)*
- U.S. Army Chem./Bio. Defense Program's Joint Effects Model (JEM) *(Kathy Houshmand, Vladimir Kogan, Joint Effects Model program)*

---

[1] The overview talks are aimed at providing some basic context for the workshop participants who come from a wide variety of professional backgrounds.
[2] Additional models and operational programs may be discussed, although this session is not aimed at providing a comprehensive review of all relevant existing activities.

- Computer-Aided Management of Emergency Operations/Areal Locations of Hazardous Atmospheres (CAMEO/ALOHA) *(Mark Miller, NOAA Office of Response and Restoration)*

NOON   Lunch

1:00 P.M.   Discuss examples of transport/dispersion model application and analysis, including:

- Persian Gulf war–modeling exercises *(Tom Warner, NCAR)*
- World Trade Center disaster, modeling of smoke dispersion *(Alan Huber, EPA/ORD)*
- Chernobyl accident–modeling of continental/global transport *(James Ellis, DOE/LLNL)*
- Salt Lake City Olympics preparations *(Brian Beitler, DTRA)*
- Urban 2000/VTMX field studies *(Gerald Streit, DOE/ LANL)*

3:00 P.M.   Briefing on the Office of Federal Coordinator for Meteorology/Joint Action Group for the Selection and Evaluation of Atmospheric Transport and Diffusion models

3:30 P.M.   Divide into breakout groups for in-depth assessment of the following topics:

- Dispersion modeling capabilities, limitations, development needs (*Chair: David Bacon, SAIC; Rapporteur: David Neff, Colorado State University*)
- Observations and data assimilation for atmospheric dispersion models (*Chair: Michael Hardesty, NOAA/ETL; Rapporteur: Peggy LeMone, NCAR*)
- Information needs of emergency first-responders and other 'user' groups (*Chair: Stephen McGrail, Massachusetts Emergency Management Agency; Rapporteur: Lewis Duncan, Dartmouth College*)

Tuesday, July 23, 2002

9:00 A.M.   Plenary session: preliminary breakout group reports and feedback from participants

10:00 A.M.   Continue breakout group discussions, with each group attempting to draft a statement to summarize findings and priorities for R&D

NOON   Lunch

1:00 P.M.   Final break-out group session

2:15 P.M.   Break

2:30 P.M.   Final plenary session: summary reports from breakout groups; closing discussion to integrate the issues raised by each group.

Participants

Committee members:

Robert Serafin (chair), National Center for Atmospheric Research
Lewis Duncan, School of Engineering, Dartmouth College
Eric Barron, Pennsylvania State University, College of Earth and Mineral Sciences
Howard Bluestein, University of Oklahoma, Department of Meteorology
Steven Clifford, University of Colorado; CIRES
Margaret LeMone, National Center for Atmospheric Research
Thomas Warner, University of Colorado, Program on Atmospheric and Oceanic Sciences
Karl Turekian, Department of Geophysics, Yale University
Gene Pfeffer, Orbital Sciences Corporation
William Odom, Hudson Institute, National Security Studies and Yale University
David Neff, Colorado State University, Department of Civil Engineering
John Wyngaard, Pennsylvania State University

Guests at July 22–23 workshop:

David Bacon, SAIC
Martin Bagley, Defense Threat Reduction Agency
Brian Beitler, Defense Threat Reduction Agency
Yu-Han Chen, MIT Department of Earth, Atmospheric, and Planetary Sciences
Walter Dabberdt, Vaisala Inc.
Paula Davidson, NOAA National Weather Service
Frances Edwards-Winslow, San Jose Emergency Preparedness Office
James Ellis, LLNL National Atmospheric Release Advisory Capability
Steve Hanna, George Mason University
Michael Hardesty, NOAA/ETL
Paul Hirschberg, NOAA National Weather Service
Kathy Houshmand, DOD Chem/Bio Defense Program (Joint Effects Model program)
Alan Huber, EPA/ORD (and NOAA/ARL)
Vladimir Kogan, Battelle (DOD Joint Effects Model program)
Donald Lucas, MIT Department of Earth, Atmospheric, and Planetary Sciences
Stephen McGrail, Massachusetts Emergency Management Agency
Mark Miller, NOAA Office of Response and Restoration

Debra Payton, NOAA Office of Response and Restoration
Jennifer Reichert, DOE Chemical-Biological National Security Program
David Roberts, Mitretek Systems
Jack Settelmaier, NOAA National Weather Service, Southern Region
John Sorensen, Oak Ridge National Laboratory
Gerald Streit, Los Alamos National Laboratory (DOE/CNBP Modeling and Prediction Project)
Gayle Sugiyama, LLNL National Atmospheric Release Advisory Capability
Samuel Williamson, Office of the Federal Coordinator for Meteorology

Guests at May 8–9 planning meeting:

Martin Bagley, Defense Threat Reduction Agency
Warren Bowen, Technical Support Working Group
Charles Hess, Federal Emergency Management Agency
Michael Lowder, Federal Emergency Management Agency
Bob Lyons, U.S. Army Soldier and Biological Chemical Command
Duncan McGill, Defense Threat Reduction Agency
Lew Podolske, White House Office of Homeland Security
David Rogers, NOAA Office of Oceanic and Atmospheric Research
Donald Wernly, NOAA National Weather Service

B

Overview of Atmospheric Transport and Dispersion Modeling

*Summary of a presentation by Steven Hanna,
George Mason University/Harvard School of Public Health*

An overview is given of the history and the current status of atmospheric transport and dispersion models applied to C/B/N releases. The discussion includes questions being asked of models, history and types of models, links to meteorological inputs, evaluations with field data, uncertainties, and future systems and research needs. Models are being applied in real time, in historical mode, and in planning mode to address the following types of concerns: In real-time, for a known C/B/N release, what areas should be evacuated or other precautions taken? Alternatively, for an unknown C/B/N release but with observed concentrations, what are the location and magnitude of the release(s)? For historical analysis, what was the dose for past C/B/N releases (e.g., Khamisiyah, Bhopal, World War I)? For planning analysis, what are the typical impacts of expected C/B/N release scenarios?

Experience shows that transport and dispersion research is driven by major events or step-changes rather than long-term planning. Examples of major events are the use of CB agents in World Wars I and II, the nuclear tests of the 1950s, the 1968 Clean Air Act and its 1990 amendments passed by the U.S. Congress, the discovery of acid lakes in the 1970s, the discovery of the ozone hole in the 1980s, the Bhopal chemical accident, the Chernobyl nuclear plant accident, the Gulf war, the Japanese subway chemical agent release, and the September 11, 2001, terrorist attacks.

BRIEF HISTORY OF TRANSPORT AND DISPERSION RESEARCH

The fundamental problem in any transport and dispersion exercise is that, no matter what model is used, the turbulence must somehow be parameterized. This has been a central theme of research over the past 80 years, beginning with Richardson and Taylor's fundamental studies. Transport and dispersion model research was funded by C/B/N concerns for several decades (e.g., the Pasquill and Calder studies in the 1940s, 1950s, and 1960s, and the Porton Down and Prairie Grass field experiments in the 1950s). There were extensive classified studies in the United States, since there was a C/B/N offensive program through the Vietnam War. Large field experiments were conducted in many types of geographic locations, such as urban areas (Fort Wayne) and coastal zones (Cape Canaveral and Vandenburgh Air Force Base). At the Department of Energy national labs and NOAA, research was carried out in the 1950s and 1960s on models for nuclear releases, fallout, and source estimation.

Over the past 20–30 years, as a result of the Clean Air Act, the research emphasis switched to EPA pollutants (e.g., SO_2) and concerns (e.g., industrial point sources, mobile sources, acid rain, regional ozone precursors, particles and toxics). Many large EPA field experiments (e.g., the St. Louis Regional Air Pollution Study and the Complex Terrain Tracer Studies) took place, and model development efforts were conducted, leading to—for example—the Models-3 regional modeling system and the AERMOD short-range model. Many urban- to regional-scale field experiments have addressed the ozone issue and, more recently, fine particles and potentially toxic chemicals. The past five years have seen a switch back to DOD and DOE, with most of the new model development and the new field experiments being supported with C/B/N concerns in mind.

The types of transport and dispersion models have evolved over the past 50–60 years, beginning with the analytical models (Gaussian, similarity, K) or nomograms used through the 1960s. In the 1970s, the focus switched to computer solutions of Gaussian plumes or of three-dimensional grid models involving the eddy diffusivity, K. The 1980s saw the development of Lagrangian puff models and one-dimensional time-dependent slab models, as well as improvement of three-dimensional Eulerian models (but with few grid nodes). Gaussian models were adapted to account for Monin-Obukhov and convective similarity, and advances were made in large eddy simulations and concentration fluctuations. In the 1990s, there were great advances in three-dimensional Eulerian models linked with numerical weather prediction models (e.g., the EPA's Models-3 system), and algorithms were improved in Gaussian-Lagrangian-puff models. So far in the 2000s, we have seen an increase in studies with CFD models, in linked emissions-meteorology-dispersion-exposure-risk systems, and in improved algorithms in Gaussian-plume models for building downwash and for concentration fluctuations.

There always have been strong links between meteorology and transport and dispersion models. Early models used a single meteorological monitor for input (e.g., NWS airport site or on-site tower). The 1970s and 1980s saw the addition of diagnostic meteorological models, which interpolate among several observing sites and add a mass conservation constraint (e.g., Lawrence Livermore National Laboratory [LLNL] MATTHEW, EPA CALMET). In the 1990s, methods were devised to accommodate NWP model outputs (although the grid was coarse and the NWP model could not be run in real time). The 2000s have seen improved grid resolution of NWP models and improved computer speed, which have allowed real-time linked NWP and dispersion models (e.g., RAMS or Eta with HYSPLIT, MM5 with CMAQ as part of Models-3, COAMPS with NARAC).

Examples of current C/B/N models include HYSPLIT and CAMEO/ALOHA from NOAA, NARAC from DOE/LLNL, HPAC from the Defense Threat Reduction Agency (DTRA), VLSTRACK from the Navy, MIDAS-AT from the Marines, the Joint Effects Model (JEM), the CATS-JACE model being developed by many agencies, and CFD models being experimented with by many groups.

Emergency response models have been needed at all times, and some examples include the Air Force's OBDG and AFTOX models from the 1960s and 1970s, the proprietary SAFER model system (including on-site meteorological instruments, dedicated computers, training, and automatic alarms) sold to hundreds of chemical plants in the 1980s, the DOE LLNL MATTHEW-ADPIC system (which was originally designed for nuclear facilities and recently has been transformed into ADAPT-LODI—part of NARAC—for C/B/N releases), the NOAA CAMEO/ALOHA system in wide use by fire departments and first responders to chemical accidents, DTRA's HPAC model and the Navy's VLSTRACK model for military applications, and NOAA's Eta-HYSPLIT model system for general purposes.

BRIEF HISTORY OF FIELD EXPERIMENTS AND MODEL EVALUATION

There has been a long history of evaluations of models with field observations. Prior to 1980, the most useful tracer experiment was the 1956 Prairie Grass study of short-range dispersion from continuous near-ground releases over flat terrain. Similar experiments took place over flat terrain as well as some urban field studies, such as the Fort Wayne study. All of these early studies were sponsored by DOD with C/B/N scenarios in mind. In the 1980s, EPA, DOE, and industrial groups such as EPRI sponsored several complex terrain field studies, some mesoscale to regional tracer experiments (e.g., CAPTEX and ANATEX), a few extensive tall stack studies (Kincaid, Bull Run, Indianapolis), and regional acid rain field experiments. In the 1990s, EPA interest focused on regional ozone studies; a few DOD mesoscale tracer studies took place such as DP26 and OLAD; and DTRA sponsored the Phase I study of ensembles of puffs. The past two years have seen an emphasis on DOD and DOE studies of releases in urban areas and obstacle arrays (e.g., MUST, Salt Lake City URBAN 2000, planned OKC-2003).

Evaluations of air quality models usually involve statistical methods such as the BOOT and ASTM software. It is found that a "good model" has a relative mean bias of about 20 or 30 percent and a scatter (normalized root-mean-square error) of a factor of 2. Most air quality models predict the ensemble mean value and not the fluctuations. An exception is HPAC, which also predicts fluctuations using standard methods from the literature. Because of the relatively large uncertainty in model predictions, the question arises of how we should inform emergency responders and other decision makers of uncertainties and of the need to consider probabilistic predictions. The study of model sensitivity and uncertainty is an expanding research area, involving methods such as probabilistic Monte Carlo uncertainty analysis.

EXPECTATIONS OF FUTURE RESEARCH

Future systems are expected to involve real-time linked source emissions modules, meteorological modules, transport and dispersion modules, and exposure and risk modules. There is a need for efficiently communicating data and model predictions across large distances (e.g., from a modeling center to a battlefield or an emergency location). Much more work is anticipated on inverse modeling or source-finding, where observations are used to triangulate to identify the location and magnitude of a release. The accelerated studies of CFD models should produce data sets for analysis and parameterization. Research needs also include better parameterizations of mean flow vectors and turbulence in the lowest 2 km for all time periods and surface types, improved methods of real-time modeling using limited inputs, development of criteria for the best expected model agreement with observations, and optimization of methods to use new remote data systems.

C

Meteorological Observing Systems for Tracking and Modeling C/B/N Plumes

Summary of a presentation by Walter F. Dabberdt, Vaisala Inc.

Meteorological observations play a critically important role in tracking and predicting the dispersion of gases and particles in the atmosphere. Depending on which variables are characterized (e.g., transport, diffusion, stability, deposition, plume rise), a wide range of meteorological parameters must be quantified. These can include wind speed and direction, temperature, humidity, precipitation type and intensity, mixing height, turbulence, and energy fluxes. Table C.1 summarizes the measurement requirements according to dispersion and meteorological variables. The specific variables that must be measured are a function of the algorithms and parameterizations used in the dispersion model. Because of their variability with height in the boundary layer, vertical profiles are important in addition to the more common practice of making meteorological measurements at or near the ground surface (see Lenschow, 1986, for a comprehensive discussion of atmospheric measurements in the planetary boundary layer). In the same way, spatial variability of the dispersion variables may necessitate multiple observing sites, model parameterizations, or judicious combinations of measurements and modeling. The following is a brief overview of the types of instrumentation that can be used to obtain the various meteorological observations. The primary focus is on measurement devices that are readily available from commercial sources, but some of the more promising research systems and concepts also are discussed.

IN SITU MEASUREMENTS

Meteorological towers (typically 6 to 10 m tall) are used widely as platforms for collecting in situ "surface" observations of wind, turbulence, temperature, and humidity. Mechanical wind sensors (bivanes, propeller vanes, etc.) have been used for decades, and their performance has improved steadily over this time. Sonic anemometers have come into widespread operational use over the past few years, having overcome earlier limitations, such as water-sensitive transducers, exposure characteristics, and price. Temperature can be measured to acceptable accuracy and precision by any of several different methods (e.g., resistance, capacitance), provided the probe is well shielded from solar insolation and properly ventilated. The vertical temperature gradient over the height of the tower is an important measurement for determining atmospheric stability and estimating turbulence. Typically, temperature gradients are measured using thermocouples or platinum resistance thermometers. The humidity or water vapor mixing ratio is a more difficult measurement, but it still can be made with acceptable accuracy and precision. The two most common methods are thin-film capacitance sensors and dewpoint measuring devices. Though less

TABLE C.1 Candidate Meteorological Observing Systems.

Dispersion Variables	Meteorological Variables (not all required; algorithm dependent)	Candidate Measurement Systems
Transport	Three-dimensional fields of wind speed and wind direction	Profiles; Doppler weather radar; RAOBs[a]; mesonets; aircraft; tethersonde; Doppler lidar
Diffusion	Turbulence; wind speed variance; wind direction variance; stability; lapse rate; mixing height; surface roughness	3D sonic anemometers; cup and vane anemometers; RAOBs; profiles; RASS; scanning microwave radiometer (maybe); tethersonde
Stability	Temperature gradient; heat flux; cloud cover; insolation or net radiation	Towers; ceilometers; profiler-RASS[b]; RAOBs aircraft; tethersonde; net radiometers; pyranometers; pyrgeometers
Deposition, wet	Precipitation rate; phase; size distribution	Weather radar (polarimetric); cloud radar; profilers
Deposition, dry	Turbulence; surface roughness	See turbulence
Plume rise	Wind speed; temperature profile; mixing height; stability	Profilers/RASS; RAOBs; lidar; ceilometer; tethersonde; aircraft

[a] RAOB stands for radiosonde observation.
[b] RASS stands for radio acoustic sounding system

common, meteorological towers can also be instrumented to measure heat and radiative fluxes and a number of other relevant meteorological and chemical variables.

For in situ upper-air measurements, balloon-borne radiosondes commonly are used on an operational basis. Radiosondes have in situ sensors that measure temperature, humidity, and pressure, while winds are measured using either of two general methods. One wind-finding method uses an onboard navigation aid receiver to measure the movement or change in location of the sonde. The second method tracks the flight of the radiosonde from the ground using radar or radio direction-finding equipment. Radiosondes are launched twice daily from 100 locations in the United States (992 locations worldwide in 1999). The typical ascent rate is 5 ms^{-1} and raw data are obtained every 1–6, seconds depending on the radiosonde type and manufacturer.

REMOTE SENSING

Remote sensing techniques are finding increasing use as an operational method to obtain vertical (and horizontal) profiles in the troposphere.

Radar wind profilers transmit short pulses of radio-frequency energy, which are scattered by clear-air atmospheric inhomogeneities and also by hydrometeors to produce a spectrum of Doppler velocities. There are numerous types of radar wind profilers available, and they can provide coverage ranging from near the surface to the lower troposphere to the lower stratosphere (depending on their radio frequency). The most commonly used measurement principle is Doppler beam swinging, which involves alternating the radar beam direction and measuring the

Doppler shift as a function of range (height) in each of several directions (pointing angles). The ambient vector velocity is then retrieved from the radial velocities along each pointing angle. Another method, called spaced-antenna profiling, transmits a single vertically directed radar beam and measures the phase relationships of the returned signal at multiple, adjacent antenna locations to retrieve the vector wind profile. Radar wind profilers provide the benefits of continuous unattended operation with high temporal resolution (5 minutes for UHF systems). Height resolution is 60–75 m with minimum heights of about 150 m; maximum height depends on atmospheric humidity and turbulence, and is typically 3–5 km for commercial UHF profilers. The lack of a dedicated UHF profiler frequency in the United States and growing commercial pressure by telecommunications providers for access to the commonly used profiler bands are concerns that require immediate attention.

Profiling of the lowest 150 m of the boundary layer is important, especially during nocturnal periods when the mixed-layer depth may be 50 m or less. So-called minisodars (profilers that use sound waves rather than radio waves) can provide the minimum range and resolution required, and they are a particularly useful complement to radar wind profilers. Unfortunately, sodars are inherently noisy (an audible signal is transmitted every few seconds) and, thus, encounter significant human resistance, especially in urban areas. Conversely, ambient noise can also impact sodar performance.

Meteorological radars and lidar (light detection and ranging) are two additional remote sensing systems useful for wind and other measurements important for dispersion and deposition. Operational Doppler meteorological radars transmit at wavelengths of 3, 5, and 10 cm; all three wavelengths can measure the radial velocity of hydrometeors, while the longer-wavelength systems can also measure clear-air velocities out to a few kilometers. Meteorological radars are especially valuable for quantifying wet deposition because of their ability to detect precipitation and estimate rain rates with reasonable accuracy over a wide area. Wind profiling radars also can detect and identify precipitation, but they yield only a single vertical profile, whereas meteorological radars can provide volumetric distributions over wide areas. Multiparameter radars transmit and measure returned signals from both horizontally and vertically polarized beams, enabling them to differentiate precipitation type (e.g., rain, snow, hail) and, thus, better estimate precipitation rates. The National Weather Service has plans to upgrade its WSR-88D weather radars to include this capability beginning around 2005. The Next Generation Weather Radar system (NEXRAD; see NRC, 1995) comprises approximately 160 WSR-88D sites throughout the United States and selected overseas locations. Figure C.1 shows NEXRAD coverage above 3 km for the contiguous United States. A limitation of NEXRAD for dispersion applications is its limited area of coverage in the lower troposphere due to Earth's curvature, blockage by obstacles, and the 0.5-degree minimum elevation angle. Networks of smaller but more densely spaced radars are being considered to complement NEXRAD and overcome these limitations (NRC, 2002).

Lidar systems emit pulses of energy at wavelengths that can vary from ultraviolet to visible to near-IR depending on the particular device. Light is scattered back from the atmosphere by particulate matter (and hydrometeors), which can serve as tracers of atmospheric mixing in the boundary layer. This enables simple backscatter lidars to estimate mixing depth, especially during unstable atmospheric conditions when there is turbulent mixing and the particulates are well mixed below the capping inversion layer. Nocturnal estimates by lidar provide higher signal-to-noise ratios but are less definitive because of uncertainties associated with "residual" particulates aloft—the result of earlier convective mixing. Ceilometers are backscatter lidars that have been demonstrated to be useful for measuring clear-air particulate profiles in and above both the daytime and the nocturnal boundary layer with 15-m height resolution and 15-m minimum range; the minimum sampling period is 15 seconds. Doppler lidars measure the range-resolved radial velocity with high resolution. For existing commercial systems, wind resolution is 0.5 ms^{-1} over range intervals of 5–50 minutes. Maximum range is a function of averaging time and can extend

FIGURE C.1 Composite WSR-88D coverage at 3 km above site level for the contiguous United States and the locations of the NWS and DOD radar sites. Courtesy of SRI International (2003).

to 16 km in clear air with 10-minute averaging (10 km with 5-minute averaging). All lidars, however, are range limited in the presence of intervening clouds.

To summarize, operational radio-, acoustic- and optical-frequency profilers provide critical atmospheric measurements needed to support dispersion and deposition modeling. Each can provide vertical profiles of wind speed, wind direction, and turbulence (derived from spectral width data), and they also are able to estimate the depth of the mixed layer(s). A comprehensive intercomparison study by Seibert et al. (2000) showed positive results at estimating mixing height from radar wind profiler, sodar, and lidar data against in situ sounding data. Bianco and Wilczak (2002) have explored the simultaneous use of data from multiple profilers using a fuzzy logic analysis scheme.

Research lidar systems offer capabilities beyond those currently available from commercial suppliers, although they tend to be more expensive and require significant human expertise to operate. However, both limitations could be minimized or eliminated in the presence of significant demand for operational systems. The research community operates two types of Doppler lidars. One is a "long-range" instrument that can sense out beyond 20 km in dry conditions and, typically, to about 15 km with higher ambient humidities. These systems are ideal for measuring flow in complex terrain, such as canyon outflows, downslope winds, and flow around barriers. The other type of research Doppler lidar is a "high-resolution," boundary layer focused lidar. These lidars have much lower pulse energy and higher pulse rates, and they are designed to probe fine-scale structure in the planetary boundary layer. These systems are typically operated in either a vertically pointing mode (for probing the convective boundary layer) or a scanning mode (stratified boundary layer), and they measure vertical velocity, vertical velocity variance, high-resolution horizontal wind profiles, and horizontal velocity variance to identify turbulent layers.

In addition to measuring winds, research lidar technology also makes it possible to obtain profiles of atmospheric properties (e.g., temperature and density) and constituents (e.g., H_2O, O_3, SO_2). Lidar sensing methods employ a wide variety of optical phenomena including elastic scattering from molecules (Rayleigh scattering) and particles (Mie scattering) where the transmitted wavelength does not change; inelastic molecular (Raman) scattering or fluorescence where the wavelength is shifted according to the type of molecule; and differential absorption where molecules absorb differentially at slightly different transmitted wavelengths.

Profiling temperature in the boundary layer and through the troposphere is also very important, especially when turbulence profiles are unavailable. Techniques for obtaining high-resolution, time-continuous temperature profiles are less well developed than those for winds and mixing height. Radiosondes are an important source of profile data for temperature but have the disadvantage of being instantaneous measurements that are available only infrequently. Radar wind profilers can measure the vertical profile of virtual temperature when configured to operate as a RASS. An acoustic source is used in RASS systems to emit intermittent sound pulses whose speed through the atmosphere is tracked by the radar wind profiler; the temperature is retrieved from the speed-of-sound measurements, which are proportional to virtual temperature. The maximum height resolution of RASS temperature profiles is 60 m and maximum range is typically 1–2 km. As with sodar, noise is a nuisance factor that limits RASS deployment in populated areas. Passive multiple-frequency, microwave radiometers have been used in research as a means to retrieve temperature profiles over deep layers of the atmosphere. Their height resolution is limited and decreases rapidly with height above the ground (Martner et al., 1992). More recently, passive single-frequency scanning microwave radiometers have been introduced; they scan in elevation and use inversion techniques to retrieve temperature profiles in the lowest 600–1000 m of the atmosphere, with a reported height resolution of 50 m. Early results are encouraging but not yet definitive.

RAPID RESPONSE MEASUREMENTS

In the context of a terrorist attack, the time, location, and nature of the source term are not known in advance and may not be known with great specificity in the minutes to hours after an attack. As a consequence, fixed meteorological observing systems that characterize dispersion in numerical models may need to be supplemented with a rapid-response deployable meteorological observing facility. There are a number of promising commercial measurement options for mobile and transportable systems. Candidates include the following:

- Low-altitude rocketsondes currently provide lower tropospheric soundings of temperature, pressure, and humidity; winds and other measurements could be added to these sondes.
- Tethered meteorological balloon systems can provide high-resolution fixed-level observations and profiles through the boundary layer
- Unmanned aerial vehicles represent a rapidly advancing airborne platform that could be adapted to measure all necessary meteorological variables as well as chemical, biological, and nuclear contaminants.

SERENDIPITOUS MEASUREMENTS

An equally important consideration is the status and availability of measurements from the many disparate surface meteorological observing stations already in operation. Numerous meteorological observations are made by local and regional networks that currently are not available to the National Weather Service or the broader scientific community. These systems primarily are surface weather stations that could provide the backbone of a surface mesonet capability for emergency response. They should be evaluated to ensure proper siting and performance specifications, and it is important that they be quality controlled. Incremental stations then could be added to optimize these mesonets as needed. Plans to evaluate, access, and use these data should be developed well in advance of an emergency event.

SUMMARY

In summary, providing meteorological observations to support response to C/B/N releases involves the following broad challenges:

- determining what measurements are essential and/or desirable;
- designing integrated observing and modeling systems and taking maximum advantage of synergies with other day-to-day applications (e.g., air pollution, mesoscale weather, hydrology, aviation);
- establishing dedicated, comprehensive meteorological observing systems near sensitive areas; and
- developing rapid-response meteorological (and chemical) observing systems.

REFERENCES

Bianco, L. and J.M. Wilczak. 2002. Convective boundary-layer depth: Improved measurement by Doppler radar wind profiler using fuzzy logic methods. *J. Atmos. Oceanic Technol.*

Lenschow, D.H. 1986. *Probing the Atmospheric Boundary Layer*, American Meteorological Society, Boston.

Martner, B.E., D.P. Wuertz, B.B. Stankey, R.G. Strauch, E.R. Westwater, K.S. Gage, W.L. Ecklund, C.L. Martin, and W.F. Dabberdt. 1992. An evaluation of wind profiler, RASS, and microwave radiometer performance. *Bull. Amer. Meteor. Soc.* 74:599-613.

NRC (National Research Council). 1995. *Toward a New National Weather Service— Assessment of NEXRAD Coverage and Associated Weather Services.* National Academy Press, Washington, D.C.

NRC. 2002. *Weather Radar Technology—Beyond NEXRAD.* National Academy Press, Washington, D.C.

Seibert, P., F. Beyrich, S.E. Gryning, S. Joffre, A. Rasmussen, and P. Tercier. 2000. Review and intercomparison of operational methods for the determination of mixing height. *Atmos. Environ.* 34:1001-1028.

D

Scientific and Technical Information Needs of Emergency First Responders

Summary of a presentation by Frances Edwards-Winslow, Ph.D., CEM,
City of San Jose, Office of Emergency Services

For the emergency response community, the adage "It's better to do something than nothing," is not always true, since the wrong response can be very costly and dangerous. Scientific and technical information is critical for helping first responders make sound decisions with regards to intelligence, warning, defense, and response to critical threats.

Depending upon the type of event, the first-responder community may include any of the following audiences:

- emergency management officials,
- public health officials,
- police, fire, and emergency medical services field personnel,
- hospitals, and
- non-governmental organizations that provide care and shelter for affected populations.

Scientific information plays a role in numerous decisions made by first responders in the minutes to hours following an event, including the following:

- population safety—evacuating versus sheltering in place; providing timely warnings to downwind populations; determining what kinds of public safety personnel need to be deployed in the community and what kind of personal protective equipment is required for first responders;
- hospitals—determining what personal protective equipment is needed for hospital staff; what symptoms to look for and decontamination or treatment modalities to prepare for;
- transit—routes to halt service; routes and stations needing decontamination before service resumption;
- built environment—actions necessary to protect storm drains, sanitary sewers, building basements, and so forth; and
- environmental concerns—assessing possible impacts on waterways, zoos, parks, home gardens (e.g., safety of produce).

In the case of an atmospheric release of a hazardous agent, the specific types of information needed by first responders may include:

- size, time, and location of release; characterization of plume movement and elevation; location of "hot" zones within the plume;
- effect of topography, vegetation, buildings, and so forth, on agent dispersion and deposition;
- medical information—exposure risk (LD_{50}, TLV)[1,2]; symptoms and treatment; interaction with other diseases (asthma, emphysema); and
- veterinary medicine—possible impacts on pets, wild population, and disease vectors.

There are a wide variety of events for which atmospheric modeling and observations can provide vital information to emergency responders, including:

- terrorism—airborne release of nuclear, radiological, chemical, and biological agents;
- smoke from forest, and wildland fires; and
- industrial accidents and release of hazardous chemicals (e.g., Bhopal).

---

[1] LD stands for lethal dose. LD_{50} is the amount of a material that causes the death of 50 percent of a group of test animals. The LD_{50} is one way to measure the short-term poisoning potential (acute toxicity) of a material.

[2] TLV stands for threshold limit value, which is the amount of exposure (for an eight-hour day, for fives days a week) without harmful effects.

E

Ensemble Simulations with Coupled Atmospheric Dynamic and Dispersion Models: Illustrating Uncertainties in Dosage Simulations

Summary of a presentation by Tom Warner, University of Colorado

Ensemble simulations made using a coupled atmospheric dynamic model and a probabilistic Lagrangian puff dispersion model were employed in a forensic analysis of the transport and dispersion of a toxic gas that may have been released near Al Muthanna, Iraq, during the Gulf War. The ensemble study had two objectives, the first of which was to determine the sensitivity of the calculated dosage fields to the choices that were to be made about the configuration of the atmospheric dynamic model. In this test, various choices were made for model physics representations and for the large-scale analyses that were used to construct the model's initial and boundary conditions. The second study objective was to examine the dispersion model's ability to use ensemble inputs to predict dosage probability distributions. Here, the dispersion model was used with the ensemble mean fields from the individual atmospheric dynamic model runs, including the variability in the individual wind fields, to generate dosage probabilities. These are compared with the explicit dosage probabilities derived from the individual runs of the coupled modeling system.

The atmospheric dynamic model was the Pennlyvania State–National Center for Atmospheric Research (NCAR) MM5 modeling system (Dudhia, 1993; Grell et al., 1994). The triply nested computational grids used grid increments of 3.3, 10, and 30 km, and they are shown in Figure E.1. The high-resolution grid was considered necessary because fine-scale desert landscape properties can influence the boundary layer depth, and lakes in the area have dynamic effects that should be resolved. There were two inner grids. One was centered over Al Muthanna in central Iraq (grid 3N), where dispersion simulations were required. Another was centered over Hafar Al-Batin (grid 3S), the area closest to Al Muthanna with a similarly arid climate and with surface and radiosonde data available for comparison with the simulations. The nested grids, each with 35 computational layers in the vertical, were two-way interacting during the simulation. Simulations proceeded simultaneously on both grids 3S and 3N. Because the lowest model computational layer was approximately 40 m above ground level, with increasing layer depths above, it was not possible for the model to resolve the shallow nocturnal planetary boundary layer well.

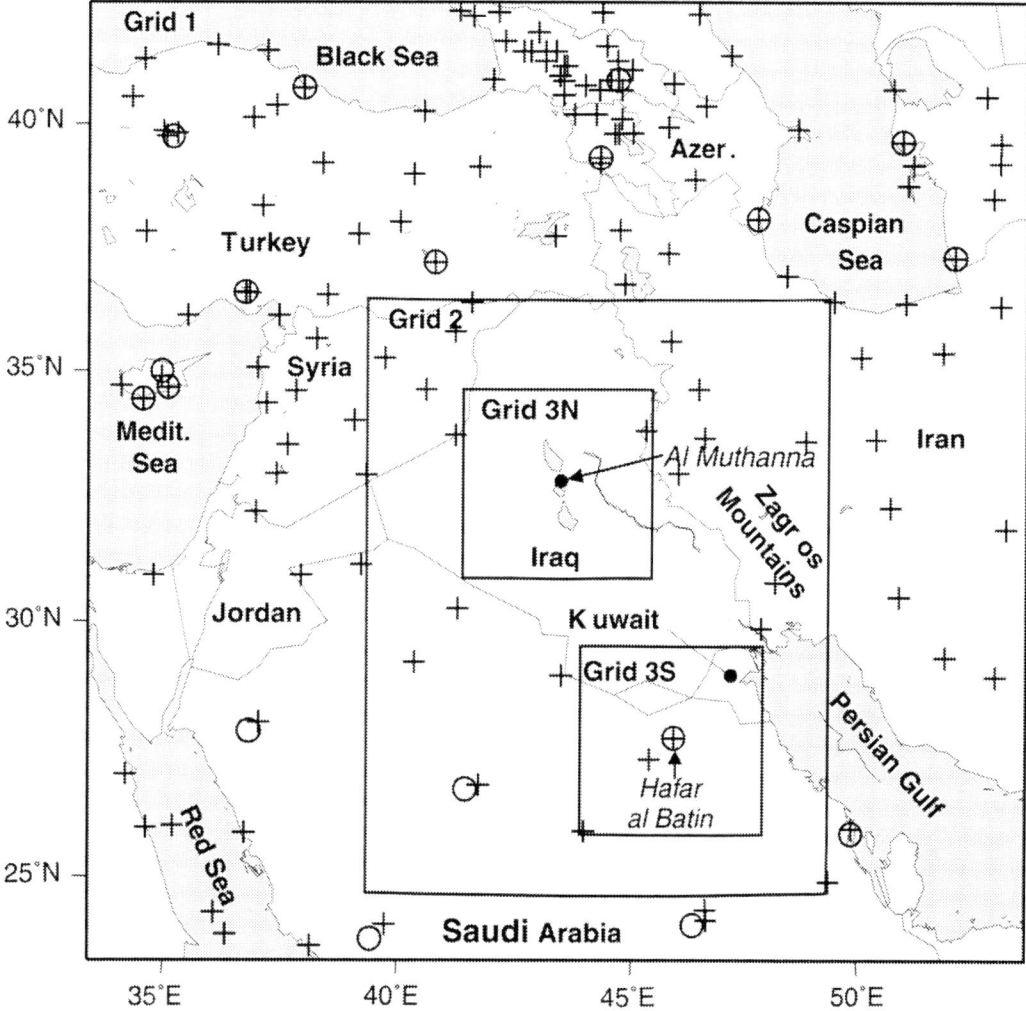

FIGURE E.1 Geographic extent of the computational grids. Grid 1 has a grid increment of 30 km, grid 2 has a grid increment of 10 km, and grids 3N and 3S have a grid increment of 3.3 km. The locations of surface (plus signs) and upper-air (circles) observations are also shown.

To create the ensemble of MM5 simulations, various options were employed for the physical process parameterizations and for the global-scale analyses that were combined with local data to generate regional atmospheric analyses. Table E.1 defines the model configurations for the various experiments performed. The MM5 model physics options used in the ensemble study included three PBL parameterizations: (1) the MRF (Medium-Range Forecast) technique used in the MRF model of the National Centers for Environmental Prediction (NCEP); (2) the turbulent kinetic energy (TKE) parameterization; and (3) the Burk–Thompson parameterization (BT). Both simple and relatively complex approaches were used for the surface energy and moisture budgets. The simpler approach employed the "slab model," in which ground temperature is calculated for a single soil layer and there is no explicit representation of vegetation effects. The more complex approach used a fairly complete land-surface model (LSM). The model initial conditions were defined by analyzing radiosonde and surface data to the model grids using a successive correction, objective analysis procedure with three different first-guess fields. The three first-guess fields were the NCEP global analysis, the European Center for Medium-Range Weather Forecasting

(ECMWF) global analysis, and the Navy Operational Global Atmospheric Prediction System (NOGAPS) analysis.

The probabilistic Lagrangian puff dispersion model used in this study was the SCIPUFF model (Sykes et al.; 1984, 1988, 1993). The acronym SCIPUFF describes two aspects of the model. First, the numerical technique employed to solve the dispersion model equations is the Gaussian-puff method in which a collection of overlapping three-dimensional puffs is used to represent an arbitrary time-dependent concentration field. The number of puffs is determined internally by the model, and it depends on such factors as the release characteristics, the size of the domain, the numerical resolution choices, and the meteorology. Second, the turbulent diffusion parameterization used in SCIPUFF is based on second-order closure theories, providing a direct relationship between measurable velocity statistics and the turbulent dispersion rates.

Plate 6 displays the SCIPUFF-calculated dosages for the different ensemble members 85 hours after release. (In most cases the gas plume had entirely exited the computational domain by this time; in the others, gas concentrations remaining on the grid were negligible.) Even though the gas moved generally to the southeast for all ensemble members, there clearly are significant differences among the solutions. In some experiments, the plume remained narrow as it traveled to the southeast. In others, the same initial movement prevailed, but the plume widened rapidly, especially toward the west. These differences result from the fact that some ensemble members carry low-level easterlies into southern and central Iraq (thus causing a westward displacement of the plume boundary), while other ensemble members do not.

One way to quantify the practical implications of the spread in the model solutions is to plot the time evolution of the area covered by the dosage above some threshold (e.g., the dosage corresponding to the "first noticeable effects" or the "general population limit"). We arbitrarily chose the lowest dosage plotted in Plate 6 for this purpose. (Note that all dosages scale exactly with the initial mass of the gas release.) Area-coverage computations were limited to the part of the grid that is within a 210-km radius of the release point. Figure E.2 shows that the areas with dosage above the threshold vary by more than a factor of four within the ensemble. In addition, it

TABLE E.1 Experimental Conditions for Each of the Ensemble-Member Simulations.

Ensemble Member Number	Large Scale Analysis Used for First Guess and Lateral Boundary Conditions	Boundary Layer Parameterizations	Surface Physics
1	ECMWF	MRF	Slab
2	NCEP	MRF	Slab
3	NOGAPS	MRF	Slab
4	ECMWF	MRF	LSM
5	NCEP	MRF	LSM
6	NOGAPS	MRF	LSM
7	ECMWF	TKE	Slab
8	NCEP	TKE	Slab
9	NOGAPS	TKE	Slab
10	ECMWF	BT	Slab
11	NCEP	BT	Slab
12	NOGAPS	BT	Slab

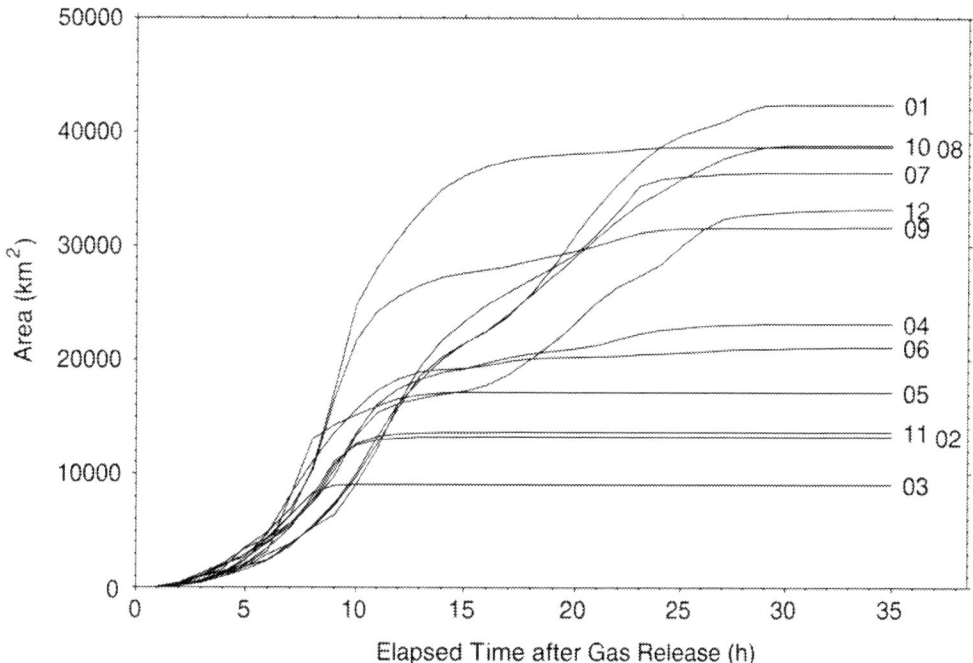

FIGURE E.2 Time evolution of the area with dosage above the threshold corresponding to the lowest value plotted in Plate 6. Area computations are limited to the part of the grid within the circle that is tangent to the sides of grid 3N, with a radius of 210 km (see arc in upper-left panel of Plate 6).

also is clear that the area coverage for the threshold dosage continues to increase out to almost 30 hours for some simulations, but for others, the area exposed reaches its maximum in as little as 8 hours.

The ensemble of dosage simulations makes it possible to calculate plots of the probability of dosages exceeding certain thresholds. Plate 7 is an isoprobability diagram for dosages exceeding 10^{-11} kg s m^{-3}. Where most of the ensemble members agree that the dosage at a location exceeds the threshold, the probability is high. Such probabilistic information is clearly much more useful to decision-makers than a single dosage simulation of unknown accuracy.

REFERENCES

Dudhia, J. 1993. A nonhydrostatic version of the Penn State/NCAR mesoscale model: Validation tests and the simulation of an Atlantic cyclone and cold front. *Mon. Wea. Rev.* 121:1493-1513.

Grell, G.A., J. Dudhia, and D.R. Stauffer. 1994. A description of the fifth generation Penn State/NCAR mesoscale model (MM5). NCAR Technical Note, NCAR/TN 398+STR, 138pp. (Available from NCAR, P.O. Box 3000, Boulder, CO 80307.)

Sykes, R.I., W.S. Lewellen, and S.F. Parker. 1984. A turbulent-transport model for concentration fluctuations and fluxes. *J. Fluid Mech.* 139:193-218.

Sykes, R.I., W.S. Lewellen, S.F. Parker, and D.S. Henn. 1988. A hierarchy of dynamic plume models incorporating uncertainty. Volume 4: Second order Closure Integrated Puff, Electric Power Research Institute, EPRI EA-6095 Volume 4, Project 1616-28. (Available from R.I. Sykes, ARAP/Titan, 50 Washington Rd., P.O. Box 2229, Princeton, NJ 08543-2229.)

Sykes, R.I., S.F. Parker, D.S. Henn, and W.S. Lewellen. 1993. Numerical simulation of ANATEX tracer data using a turbulent closure model for long-range dispersion. *J. Appl. Meteor.* 32:929-947.

Warner, T.T., R.S. Sheu, J. Bowers, R.I. Sykes, G.C. Dodd, and D.S. Henn. 2001. Ensemble simulations with coupled atmospheric dynamic and dispersion models: Illustrating uncertainties in dosage simulations. *J. Appl. Meteor.* 41:488-504.

F

Modeling Studies of the Dispersion of Smoke Plumes from the World Trade Center Fires

Summary of a presentation by Alan Huber, NOAA/ARL and EPA/ORD

The EPA and NOAA–Air Resources Laboratory had begun observational plume modeling studies in lower Manhattan, focusing on air pollution exposure assessment, in early 2001. For these studies, they developed a portable, battery-operated meteorological observing system that utilized a minisodar and a 10-m tower to obtain wind data, and they had collected several months of data from this system. They also compiled a digital model of the building topography of lower Manhattan. Following the September 11, 2001, terrorist attacks, the other applications of this work became immediately obvious.

The CALPUFF model (a Gaussian puff model) was used track the dispersion of emissions from the fires at the World Trade Center site for the period September 11 through December 8. They started with an initial assumption of the volume source, tracked the dilution of that source, and integrated the results over time to estimate possible exposure in surrounding neighborhoods. The meteorological data came from wind fields generated by CALMET (a diagnostic model) and from assimilation of surface meteorology and ARPS model data.

The CALPUFF–CALMET system was found to perform fairly well and be valuable for making forecasts in a real-time mode. However, to complement this work, EPA is developing a much more sophisticated, finer-scale CFD model and will also be carrying out wind-tunnel physical modeling studies.

The information from these dispersion modeling studies will be used to estimate the potential exposure of various populations around ground zero, providing input for epidemiological assessments of possible health impacts resulting from this exposure.

Several important lessons were learned from this work:

- It is important to do routine meteorological observations and modeling in major cities so that you can develop an understanding of the local-scale flow features before an emergency event occurs.
- Simple plume models are not sufficient for tracking dispersion in a dense urban area. Buildings and other aspects of the urban environment have a huge effect on flow and dispersion patterns.

- Meteorological data collected from many standard observational sites intentionally located in open areas (such as those based at airports) do not necessarily represent conditions occurring in nearby urban areas.

- The simple observation of which way the wind is blowing is very important for some purposes, for example, to tell people where they should be monitoring for possible exposure and health impacts.

G

Use of Atmospheric Models in Response to the Chernobyl Disaster

Summary of a presentation by James Ellis, LLNL/NARAC[1]

The Chernobyl nuclear accident occurred on April 26, 1986. The size of this release was unprecedented, releasing millions of curies of radioactive material—including iodine-131, cesium-137, and strontium-90—all of which are potentially harmful to human health.

The initial release occurred on a Friday night. By Sunday afternoon, contamination readings were picked up on workers at a nuclear power plant in Sweden. Within a few hours it had been determined that the contamination source was from a nuclear power plant to the south. By Monday, the Russians admitted that a major accident had occurred, and on the same day, LLNL/NARAC was notified by the Department of Energy to begin predicting the consequences. NARAC worked round-the-clock for two weeks, providing assistance in modeling the transport and deposition of the radioactive cloud.

NARAC utilized three different model codes for these analyses 2BPUFF, PATRIC, and MATHEW/ADPIC. The 2BPUFF model is a two-dimensional long-range transport and diffusion model used mostly for estimating the Chernobyl accident release amounts of radioactivity. The PATRIC model is a three-dimensional puff and diffusion model that had been specifically designed to treat continental and hemispheric scales. The MATHEW/ADPIC is a combined mass-consistent wind flow model and a particle-in-cell dispersion model that was to calculate consequences over 200 km or less. For the previous 12-year period, real-time radiological dose assessments had been done at scales up to 200 km. For the Chernobyl event, MATHEW/ADPIC had to be rapidly modified to expand its capability to approximately a 2000-km domain. The Air Force Global Weather Center provided meteorological data for this work.

Based on samples collected around Europe, the time and strength of the release were estimated, and approximately 40 percent of the total radioactive material was estimated to have been released in the initial blast (the rest came from the ensuing fire over the next five to six days). There was a considerable amount of rain in the area, which led to "hotspots" of wet deposition across Europe. Scientists were not able to model these washout processes, because they did not have the needed meteorological precipitation data or fine-scale forecast model precipitation pro-

---

[1] Ellis emphasized that it is difficult to reconstruct the exact history of LLNL involvement in this event, since those who participated directly in this response effort are no longer at LLNL.

ducts. With today's high-resolution mesoscale forecasts, they would have been able to do a much better job of modeling the deposition patterns.

NARAC models were used to estimate air concentration and ground deposition of key radioactive elements and the corresponding exposures and potential health effects. It was estimated that iodine deposition in the United States was insignificant but that dangerous levels of ground deposition of key radioactive elements had occurred throughout Europe. The greatest risk was not from direct exposure but from exposure through the food chain. In particular, radioactive material was deposited on farmlands and on grass eaten by cattle, forcing many countries to destroy exposed milk and crops.

Aircraft measurements of radioactive material at 17,000–30,000-ft altitude above Europe, the Japan Sea, and the West Coast of the United States indicated that radioactivity from the reactor accident had gotten higher in the atmosphere than initially thought to have been possible. Based on NARAC's knowledge of the thermal energy of the blast, it did not understand how the material could have risen so high; after examining the prevailing weather patterns, NARAC surmised that convective activity in the area had driven the radioactive material up to these high altitudes in the atmosphere.

The upper-level flow reached the United States (from across the Pacific) by Day 10. LLNL modeling of this event matched well with the readings from aircraft measurements. Even with the limitations of the meteorological data and model prediction capability available in the 1986 time frame, overall agreement between ground-based and aircraft measurements and model estimates was within a factor of 2 or 3.

Since that event, there has been a lot of activity in Europe to improve the models used in response to nuclear accidents, culminating in development of the Realtime Online Decision Support System (RODOS) for nuclear emergency management (www.rodos.fzk.de/RodosHomePage). The dispersion models being used in this system and those being used by other national organizations within Europe have been improved from those originally used in the Chernobyl response. One of the strengths of RODOS has been to link these dispersion models to better atmospheric prediction models and to various dose pathway models, including sophisticated watershed models, with the objective of providing tools to the decision-maker for making well-informed decisions.

H

Preparatory Exercises at the Salt Lake City Olympics

Summary of a presentation by Brian Beitler, DTRA

At the 2002 Olympic Winter Games in Salt Lake City, Utah, DTRA played a key role in preparing for the possibility of a terrorist attack involving an atmospheric release of hazardous agents. DTRA and several other groups involved in this work operated out of a central "smart building" that was fully equipped with computing, communications, and atmospheric monitoring equipment. This building also had the capability of protecting its inhabitants in case of a nearby release.

The primary dispersion modeling system employed in this work was the HPAC (described earlier). The primary capabilities were based at DTRA headquarters in Alexandria, Virginia, but the groups planned for redundancy, with backup systems running at NCAR and Dugway Proving Ground. Meteorological data servers, which provide a critical source of input for the HPAC system, were available from three different locations (Alexandria, Virginia; Dugway Proving Grounds; and Salt Lake City, Utah [SLC]). Investigators also had continuous, real-time atmospheric monitoring in SLC throughout the games and drew upon daily SLC forecast discussions and teleconferences with the SLC National Weather Service.

The weather forecasts employed were split into two regimes:

1. A 0–12-hour forecast, generated with the MM5 model, which could be "nudged" with real-time observations from SLC's mesonet system; and
2. A 12–36-hour forecast, generated with their "expert system" in combination with high-resolution forecasts from the RAMS/OMEGA modeling system and the University of Utah's MM5.

They ran an intercomparison test of available modeling systems, simulating the release of a nerve agent from a sprayer. All of the models used to simulate the resulting plume gave slightly different answers. In a comparison to "ground truth" obtained by a local mesonet system, investigators found that an ensemble mean of all the model simulations seemed to perform better than any single model.

Several important lessons were learned from this work:

- It is valuable to have redundancy in all of the critical systems (monitoring, computing, communications, etc.).

- One of the biggest technical challenges can be dealing with communication issues (file transfer protocol limitations, firewalls, etc.).
- It is impossible to plan for every situation, so flexibility in operations is necessary. It is also important to have contingency and backup plans for disseminating dispersion model forecasts and other data products.
- It is not clear which provides a better measure of 'truth'—a high-resolution model that is regularly spaced or the actual observations that are irregularly spaced.
- There is a benefit to using high-fidelity weather data. When the local details of topography were included in the SLC forecasts (e.g., upslope and downslope flows), they produced very complex plumes.

DTRA is currently investigating simpler alternatives for generating transport and dispersion forecasts quickly. For instance, in mountainous areas such as SLC, the decision of whether you are allowed to use wood-burning stoves is based on the ventilation index, a function of boundary layer height and wind speed. A poor ventilation index means a low boundary layer and wind speed, so any release will be trapped closer to the ground. This type of simple parameterization may lend itself to "quick look" dispersion forecasting as well.

I

URBAN 2000 Overview

Summary of a presentation by Gerald Streit, DOE Los Alamos National Laboratory

The URBAN 2000 tracer and meteorological experiments were conducted during October 2000, and they provide a unique set of nighttime atmospheric dispersion data covering transport scales from individual buildings on through the urban- to the regional-scale. The URBAN 2000 researchers collaborated closely with DOE's Environmental Meteorology Program by adding building-scale through urban-scale experiments (URBAN 2000) to their regional-scale Vertical Transport and Mixing Experiments (VTMX) in the greater Salt Lake City area.

Meteorological measurement and tracer sampling instruments were installed throughout Salt Lake City and operated for most of the month of October 2000. Instruments were sited to resolve scales of motion ranging from flows around individual buildings in downtown Salt Lake City to flows throughout the urban area. The scale of the URBAN 2000 experiment was defined by an outermost 6 km arc of fixed sampler boxes and track for one of the plume-chasing vans. A five-block by five-block focus area was more heavily instrumented and the central experimental site was intensely instrumented during IOPs (Intensive Operations Period). The mobile van, gas chromatograph, IR, LLNL sonic anemometers, and all sampling instrumentation were deployed only during the IOPs.

Further mention should be made of the six NOAA vans equipped with fast-response gas chromatographs for SF_6 detection. Four of the vans did plume chasing during the IOPs roughly following 1, 2, 4, and 6 km arcs to the northwest of the release site. Two vans remained at fixed locations. During IOPs 2 and 4, Litton Industries deployed a van with a volume scanning Fourier transform infrared spectrometer. This was used relatively near the release site to map the vertical extent of the SF_6 plume. For a little less than two weeks, from October 19 at 1800 MDT to October 27 at 1100 MDT, Coherent Technologies Incorporated deployed a wind-tracer doppler lidar at a site 4 km east of downtown and approximately 400 m higher than downtown. These dates covered IOPs 8-10. This unit mapped out the radial component of the wind in three dimensions over the city and up nearby canyons.

Table I.1 gives detail about the shakedown IOP and six full-scale URBAN 2000 IOPs that were nested within the ten VTMX IOPs. Time-integrated tracer samples (nominally 5-minute to 2-hour integration times) were collected by 200 samplers located throughout the Salt Lake Basin. The sampling period extended from just before tracer release start (~2300 MDT) through the night until the next afternoon (~1300 MDT). The tracer samplers were distributed with the intent to resolve the various scales of motion being studied: 45 SF_6 samplers were located around the

TABLE I.1 Specifics about the shakedown IOP and six full-scale URBAN 2000 IOPs that were nested within the ten VTMX IOPs

IOP#[1]	Start Time[2]		SF$_6$ Source Geometry	SF$_6$ Release Start Time (MDT) (Releases Were 1-hour Duration)			Sampling End Time (MDT) (Start Day +1)	Comments
	UTC	MDT		Release1	Release 2	Release 3		
1	2-Oct-00 2200	2-Oct-00 1600	Point	0100 (2gs^{-1})	0300 (1gs^{-1})		0500	Shakedown; met[3,4], mux GC, mux IR, and 2 mobile vans deployed; no SF$_6$ box samplers and no PFTs.
2	6-Oct-00 2200	6-Oct-00 1600	Line, 1 gs^{-1}	0100	0300	0500	1300	Full met, SF$_6$, and PFT experiment, low winds.
3	7-Oct-00 2200	7-Oct-00 1600						VTMX only.
4	8-Oct-00 2200	8-Oct-00 1600	Line, 1 gs^{-1}	0100	0300	0500	1300	Full met, SF$_6$, and PFT experiment, low winds.
5	14-Oct-00 2200	14-Oct-00 1600	Line, 1 gs^{-1}	0100	0300	0500	1300	Full met, SF$_6$, and PFT experiment, low winds.
6	15-Oct-00 2200	15-Oct-00 1600						VTMX only.
7	17-Oct-00 2200	17-Oct-00 1600	Line, 1 gs^{-1}	0100	0300	0500	1300	Full met, SF$_6$, and PFT experiment, low winds.
8	19-Oct-00 2200	19-Oct-00 1600						VTMX and URBAN PFT[5,6] experiment; no met or SF$_6$.
9	21-Oct-00 0400	20-Oct-00 2200	Point, 2 gs^{-1}	2200	0000	0200	0400	Full met and SF$_6$ experiment, higher winds; no PFT.
10	25-Oct-00 2200	25-Oct-00 1600	Point, 1 gs^{-1}	0100	0300	0500	1300	Full met, SF$_6$, and PFT experiment, low to higher winds.

[1] IOP stands for intensive operation period
[2] Time to first balloon launch by VTMX.
[3] Met refers to those instruments deployed just for the IOP.
[4] Met, mux GC, and mux IR were taken down within 1 hour after the end of the final release
[5] URBAN PFT point source releases were continuous from 0100-0700 MDT
[6] URBAN utilized two different PFTs, one at each source location

downtown study buildings, 40 combined SF_6/PFT (Perfluorocarbon Tracer) samplers and 24 SF_6 samplers were located in a 5-block-square area (25 blocks) of downtown; 36 SF_6 samplers were located on three sampling arcs (2, 4, and 6 km) to the northwest of the downtown SF_6 release location; and 55 PFT samplers were located throughout the Salt Lake Basin. A total of nearly 11,000 SF_6 samples and 5,000 PFT samples were collected during the tracer experiments. In addition to the 200 tracer samplers deployed during the combined VTMX/URBAN 2000 experiments, two SF_6 analyzers were deployed by LLNL during the IOPs around the downtown study building.

A summary of meteorological instrumentation deployed for URBAN 2000 follows:

- Building scale (completely within the core block): 12 two-dimensional sonic anemometers (the five long-term locations included temperature measurements), 2 three-dimensional sonic anemometers, and 1 laser ceilometer;
- Urban scale (a five-block by five-block square): 10 portable meterological stations, 3 two-dimensional sonic anemometers (1 station included temperature), 7 three-dimensional sonic anemometers, and 1 acoustic sodar; and
- 1–6–km scale: 6 wind stations, 2 acoustic sodars, 1 radar wind profiler, 54 temperature loggers, 1 Doppler lidar.

The Pacific Northwest National Laboratory temperature loggers were sited on a north-to-south transect and on a west-to-east transect across Salt Lake City collecting 15-minute-average data for the month of October. They were located on 400 South from 1500 West to 1500 East, and on State from 1500 South to approximately 1500 North, so they crossed the urban- building-scale regimes. Some very early results for plume concentration measurements during IOP10 are shown in Plate 8.

REFERENCE

Allwine, K.J., J.H. Shinn, G.E. Streit, K.L. Clawson, and M.J. Brown. 2002. Overview of URBAN 2000: A multi-scale field study of dispersion through an urban environment. *Bull. Amer. Meteor. Soc.* 83:521-536.